Data Wrangling with SQL

A hands-on guide to manipulating, wrangling, and engineering data using SQL

Raghav Kandarpa

Shivangi Saxena

BIRMINGHAM—MUMBAI

Data Wrangling with SQL

Copyright © 2023 Packt Publishing

Group Product Manager: Kaustubh Manglurkar

Publishing Product Manager: Arindam Majumder

Senior Editor: Sushma Reddy

Technical Editor: Kavyashree K. S.

Copy Editor: Safis Editing

Project Coordinator: Hemangi Lotlikar

Proofreader: Safis Editing

Indexer: Hemangini Bari

Production Designer: Shyam Sundar Korumilli

Marketing Coordinator: Nivedita Singh

First published: July 2023

Production reference: 1280723

Published by Packt Publishing Ltd.
Grosvenor House
11 St Paul's Square
Birmingham
B3 1R

ISBN 978-1-83763-002-8

www.packtpub.com

To our adorable nephew, who filled our writing journey with laughter, joy, and endless inspiration. Though you may be just two years old, your presence has brought an abundance of love and light into our lives. This book is dedicated to you, a reminder of the cherished moments we shared as we crafted its pages together. May it serve as a symbol of our bond, and may it inspire you to embrace your own passions and embark on remarkable adventures.

Acknowledgements

Writing a book is a journey of dedication and perseverance, and we are immensely grateful for the support and contributions of numerous individuals who have made this endeavor possible.

First and foremost, we would like to express our deepest gratitude to each other. The process of writing this book was not only a professional collaboration but also a personal journey for us. Amid the demands of deadlines and research, we took a significant step in our lives and got married. Our love and partnership have been the driving force behind this book, and we are grateful for the incredible bond that sustains us through both personal and professional challenges.

We would like to extend our heartfelt appreciation to our families, whose unwavering love and support have been a constant source of strength. Your belief in us and the encouragement you provided throughout this journey have been instrumental in our success. We are grateful for your understanding during the moments when our focus was fully dedicated to the completion of this project.

To our editor and the entire publishing team, we extend our sincere thanks. Your guidance, patience, and professionalism have been invaluable throughout the book-writing process. We appreciate your belief in our work and the opportunity to collaborate with such a talented group of individuals.

We would also like to acknowledge the reviewers and technical experts who generously shared their knowledge and insights. Your feedback and suggestions have played a vital role in refining the content of this book and ensuring its accuracy and relevance. We are grateful for the time and expertise you dedicated to reviewing our work.

Lastly, we want to express our deepest appreciation to the readers who embark on this journey with us. Your curiosity and enthusiasm for data wrangling using SQL inspire us, and we hope that this book serves as a valuable resource in your own endeavors. Thank you for being part of our journey and for your support and engagement.

Writing this book has been an extraordinary experience, blending personal and professional milestones. We are humbled by the love and support we have received from our loved ones and the publishing community. Thank you all for standing by our side, celebrating our marriage, and contributing to the creation of this book.

With heartfelt gratitude,

Raghav and Shivangi

Dedication

This book is dedicated to our families, whose unwavering love and support have been the foundation of our lives. Your belief in us and the encouragement you have given throughout this journey have been instrumental in our success.

To our parents, who instilled in us the values of hard work, perseverance, and the pursuit of knowledge; we are forever grateful. Your guidance and sacrifices have shaped us into the individuals we are today, and this book is a testament to your unwavering belief in our abilities.

To our siblings, whose unwavering support and camaraderie have brought us joy and inspiration; thank you for always being there, cheering us on, and offering a listening ear when we needed it most.

To our extended families, relatives, and friends who have celebrated our achievements, offered words of encouragement, and lent a helping hand along the way; your presence in our lives has brought warmth and happiness. Your love and support have meant the world to us.

Above all, this dedication is for each other. Our love and partnership have been the driving force behind this book. Through the late nights, the endless revisions, and the challenges we faced, we found strength and solace in each other. Your unwavering belief in our shared vision and the constant support you provide are the cornerstones of our journey.

May this book serve as a tribute to the love, support, and dedication of our families and the power of partnership. We dedicate this work to all those who have played a part, big or small, in shaping our lives and making this book possible.

With all our love and gratitude,

Raghav and Shivangi

Contributors

About the authors

Raghav Kandarpa is a skilled author and data analyst passionate about extracting insights from complex datasets. Originally from India, he pursued his academic and professional goals in data analysis in the United States. With a master's degree in business analysis, specializing in business intelligence and analytics, he possesses expertise in data-driven decision-making and advanced analytical techniques. Currently a lead data science analyst at a top Fortune 500 finance company, Raghav leverages data to drive strategic initiatives, delivering actionable recommendations and optimizing business operations. Alongside his professional achievements, he is dedicated to empowering aspiring data analysts through practical writing that bridges the gap between theory and practice. Raghav's commitment to innovation, data-driven decision-making, and knowledge sharing continues to shape the future of data analysis in finance and beyond.

I want to thank the people who have been close to me and supported me, especially my parents and my brother.

Shivangi Saxena, a highly accomplished data analyst and Business Intelligence Engineer with over six years of experience, relocated from India to the United States to pursue a master's degree in information technology and management, specializing in business intelligence and analytics. She quickly became a valuable asset in the industry, leveraging data to drive strategic business decisions with her strong analytical mindset and passion for technology. Currently working with a top Fortune 500 company, Shivangi excels in data analysis, visualization, and business intelligence, making her a trusted advisor. Her ability to derive meaningful insights from complex datasets and translate them into actionable strategies sets her apart. Committed to empowering others, Shivangi has written this book to help aspiring data analysts enhance their skills. Her writing combines theory and practical application to offer valuable guidance. Additionally, Shivangi actively mentors aspiring data professionals, particularly women, encouraging them to pursue their dreams in the field. Her journey reflects her resilience, determination, and passion for data analysis, as she continues to inspire others and shape the future of the industry with her expertise and thought leadership.

I would like to thank my loving and patient parents and siblings for their continued support, patience, and encouragement throughout the long process of writing this book.

About the reviewers

Charles Mendelson is a Seattle-based software engineer who has worked at virtually every stage of company, from early-stage start-ups with no infrastructure to century-old operations full of legacy systems. He has a master's degree from Harvard University where he studied psychometrics (how you measure things in psychology) at the Harvard Extension School. In 2022, he was named one of the top 25 data-engineering influencers on LinkedIn by Databand.

He is also on the instructional staff in the Python Certificate Program at the University of Washington's School of Professional and Continuing Education, where he helps adult learners develop coding skills.

At his primary employer, PitchBook Data, he has built a training program called **Women in Engineering (WinE)** to help women early in their careers develop technical skills, particularly in Python, SQL, and Excel.

In 2017, he was published in the Harvard Business Review, and since 2020, he has been a contributing writer for **Towards Data Science**, where he primarily writes articles on structuring and organizing data.

Jessica Ginesta is an accomplished professional with a diverse background and extensive experience in pre-sales and customer success roles. With over seven years of expertise in the data space, she possesses a deep understanding of SQL, analytics, and AI technologies. Her skill set includes proficiency in enterprise software, big data/analytics, data engineering, and data science.

During her tenure at Databricks, Jessica has excelled in her role as a senior solutions architect. Jessica's exceptional proficiency in SQL has allowed her to leverage data effectively, providing valuable insights to clients. Her expertise in analytics and AI enables her to develop and deliver innovative solutions that drive business growth and address critical challenges. Jessica's strategic mindset and ability to communicate complex solutions in a clear and concise manner have made her a trusted advisor to clients and an asset to her team.

Prior to joining Databricks, Jessica held positions in renowned organizations where she played a pivotal role in utilizing SQL, analytics, and AI to deliver impactful results. Her passion for leveraging data to drive decision-making and her strong track record of success have earned her recognition as a leader in the field.

Mohammed Kamil Khan is currently a graduate student at the **University of Houston-Downtown (UHD)**, majoring in data analytics. He has accumulated 1.5 years of experience in various analytics-related positions. He is now working as a research assistant at UHD, engaged in a study funded by a grant from the **National Institutes of Health (NIH)**. With an unwavering passion for democratizing knowledge, Kamil strives to make complex concepts accessible to all. Moreover, Kamil's commitment to sharing his expertise led him to publish tutorial-based articles on platforms including DigitalOcean, Open Source For You magazine, and Red Hat's `opensource.com`. These articles explore a diverse range of topics, such as pandas DataFrames, API data extraction, SQL queries, and the Django REST framework for API development.

Table of Contents

Part 1: Data Wrangling Introduction

1

Database Introduction

2

Data Profiling and Preparation before Data Wrangling

Part 2: Data Wrangling Techniques Using SQL

3

Data Wrangling on String Data Types   61

4

Data Wrangling on the DATE Data Type 89

5

Handling NULL Values 125

6

Pivoting Data Using SQL 141

Part 3: SQL Subqueries, Aggregate And Window Functions

7

8

9

Part 4: Optimizing Query Performance

10

Optimizing Query Performance   255

In the next chapter, we will learn about descriptive statistics using SQL, which will provide us with insights into the distribution, central tendency, and variability of data, which can, in turn, help us identify outliers and anomalies. Common SQL functions and statements used for descriptive statistics include `COUNT`, `AVG`, `MIN`, `MAX`, and `GROUP BY`. By using SQL to analyze data, researchers and analysts can efficiently extract and summarize information from large datasets.   267

Part 5: Data Science And Wrangling

11

Descriptive Statistics with SQL   271

In the next chapter, we will learn how SQL can be used for time series analysis.   282

12

Time Series with SQL   283

13

Outlier Detection   301

Preface

Welcome to *Data Wrangling with SQL*, a comprehensive guide that equips you with essential skills to efficiently manipulate and prepare data for analysis using SQL. In today's digital age, harnessing the power of data has become crucial for individuals and organizations across industries. Data wrangling, also known as data preprocessing or data munging, involves transforming raw data into a clean, structured format for easy analysis. It encompasses handling missing values, removing outliers, merging datasets, and reshaping data to meet analysis requirements. Effective data wrangling lays the foundation for accurate and meaningful insights, enabling informed decision-making.

This book aims to demystify the art of data wrangling using SQL, a powerful language for data manipulation. Whether you are a data analyst, business intelligence professional, or data enthusiast, this guide provides the knowledge and skills necessary to navigate data preparation complexities. Throughout this book, we explore SQL techniques and best practices for data wrangling. Starting with basics such as selecting, filtering, and sorting data, we delve into advanced topics such as data aggregation, joins, subqueries, and data transformation functions.

While SQL's versatility can be overwhelming, this book balances theoretical concepts with practical examples and hands-on exercises. Real-world scenarios and datasets reinforce your understanding, fostering confidence in tackling data-wrangling challenges. Beyond technical skills, we discuss strategies for handling common data quality issues, ensuring integrity, and improving performance. Techniques for handling missing data, outliers, and inconsistencies are explored, alongside common pitfalls and tips for optimizing SQL queries.

Approach each chapter with curiosity and enthusiasm, embracing challenges to deepen understanding and enhance problem-solving abilities. Data wrangling is not just a process but also an art form, requiring creativity, logical thinking, and attention to detail.

This book serves as a valuable resource in your journey to becoming a proficient data wrangler. May it empower you to unlock data's full potential, uncover hidden insights, and drive meaningful impact.

Happy wrangling!

Who this book is for

Data Wrangling with SQL is intended for individuals who are interested in data analysis, data manipulation, and data preparation using the SQL language. The book caters to a diverse audience, including the following:

- **Aspiring data analysts**: Individuals who want to enter the field of data analysis and gain a solid foundation in data-wrangling techniques using SQL

- **Business intelligence professionals**: Professionals working in business intelligence roles who wish to enhance their SQL skills for effective data preparation and analysis

- **Data enthusiasts**: Individuals passionate about data and eager to acquire practical skills in data wrangling using SQL

What this book covers

Chapter 1, *Database Introduction*, is where you will discover the fundamentals of databases and their role in data wrangling, equipping you with a solid foundation to leverage SQL for efficient data manipulation and analysis.

Chapter 2, *Data Profiling and Preparation before Data Wrangling*, is where you will master the art of data profiling and preparation, empowering you to assess data quality, handle missing values, address outliers, and ensure data integrity before diving into the data-wrangling process using SQL.

Chapter 3, *Data Wrangling on String Data Types*, explores the ins and outs of manipulating and transforming string data using SQL, enabling you to clean, format, extract, and combine textual information efficiently in your data-wrangling workflows.

Chapter 4, *Data Wrangling on the DATE Data Type*, unlocks the power of SQL to handle date data effectively, covering techniques for date formatting, extraction, manipulation, and calculations, allowing you to wrangle temporal data with precision and accuracy in your analysis.

Chapter 5, *Handling NULL Values*, navigates the complexities of NULL values in datasets and teaches you SQL techniques to identify, handle, and manage null values effectively, ensuring data integrity and enabling seamless data wrangling for accurate analysis.

Chapter 6, *Pivoting Data Using SQL*, will help you master the art of transforming row-based data into a structured columnar format using SQL, enabling you to pivot and reshape data for enhanced analysis and reporting capabilities in your data-wrangling endeavors.

Chapter 7, *Subqueries and CTEs*, dives into the world of subqueries and **Common Table Expressions** (**CTEs**) in SQL, mastering the art of structuring complex queries, enhancing data-wrangling capabilities, and simplifying your data analysis workflows for optimal efficiency and clarity.

Chapter 8, *Aggregate Functions*, unleashes the power of aggregate functions in SQL, empowering you to perform powerful calculations and summarizations on your data, enabling effective data wrangling for extracting insightful statistics and metrics in your analysis workflows.

Chapter 9, SQL Window Functions, unlocks the advanced capabilities of SQL window functions, enabling you to perform complex calculations and analyses over customized subsets of data, revolutionizing your data-wrangling techniques for insightful data partitions, rankings, and aggregations.

Chapter 10, Optimizing Query Performance, helps you master the art of optimizing SQL queries, exploring techniques and strategies to enhance query performance, minimize execution time, and maximize efficiency in your data-wrangling workflows, ensuring faster and more effective data analysis.

Chapter 11, Descriptive Statistics with SQL, shows you how to harness the power of SQL to perform descriptive statistical analysis on your data, exploring SQL functions and techniques to extract key insights, summarize data distributions, and uncover patterns, enabling data wrangling for robust exploratory data analysis.

Chapter 12, Time Series with SQL, unleashes the potential of SQL for time-series analysis, exploring techniques for manipulating, aggregating, and extracting valuable insights from temporal data, empowering you to conduct effective data wrangling and uncover trends and patterns in your time series datasets.

Chapter 13, Outlier Detection, helps you master the art of identifying and handling outliers in your data using SQL, equipping you with techniques and strategies to detect, analyze, and manage outliers effectively in your data-wrangling workflows, ensuring data integrity and accurate analysis.

To get the most out of this book

To make the most of this book, readers should have a basic understanding of SQL fundamentals, including database concepts, querying, and manipulating data using SQL statements. Familiarity with relational databases and SQL syntax is beneficial.

While prior experience with data analysis or business intelligence is not mandatory, a general understanding of data analysis concepts and the importance of data preparation will be advantageous.

The book's progressive approach, starting from foundational concepts and gradually advancing to more complex topics, allows readers with varying levels of SQL expertise to benefit from the content.

Regardless of their background, readers should come with an eagerness to learn, a willingness to explore practical examples, and a desire to develop their data-wrangling skills using SQL.

By catering to both beginners and those with some SQL experience, this book aims to provide valuable insights and techniques for individuals at different stages of their data analysis journey.

Software/hardware requirements	Operating system requirements
SQL database	
MySQL	Windows or macOS
SQL Server (for some examples)	

If you are using the digital version of this book, we advise you to type the code yourself or access the code from the book's GitHub repository (a link is available in the next section). Doing so will help you avoid any potential errors related to the copying and pasting of code.

Download the example code files

You can download the example code files for this book from GitHub at `https://github.com/PacktPublishing/Data-Wrangling-with-SQL`. If there's an update to the code, it will be updated in the GitHub repository.

We also have other code bundles from our rich catalog of books and videos available at `https://github.com/PacktPublishing/`. Check them out!

Conventions used

There are a number of text conventions used throughout this book.

`Code in text`: Indicates code words in text, database table names, folder names, filenames, file extensions, pathnames, dummy URLs, user input, and Twitter handles. Here is an example: "The syntax for creating a database is – `CREATE DATABASE Database_name`."

A block of code is set as follows:

```
Create table walmart.customer_info
(
CustomerID int,
Name varchar(255),
Address varchar(255),
Email varchar(255),
Phone varchar(255)
)
```

When we wish to draw your attention to a particular part of a code block, the relevant lines or items are set in bold:

```
Select lower(Address) from customers
```

Bold: Indicates a new term, an important word, or words that you see onscreen. For instance, words in menus or dialog boxes appear in **bold**. Here is an example: "In the preceding code, the **customerID** column is defined as an integer data type"

> **Tips or important notes**
> Appear like this.

Get in touch

Feedback from our readers is always welcome.

General feedback: If you have questions about any aspect of this book, email us at `customercare@packtpub.com` and mention the book title in the subject of your message.

Errata: Although we have taken every care to ensure the accuracy of our content, mistakes do happen. If you have found a mistake in this book, we would be grateful if you would report this to us. Please visit `www.packtpub.com/support/errata` and fill in the form.

Piracy: If you come across any illegal copies of our works in any form on the internet, we would be grateful if you would provide us with the location address or website name. Please contact us at `copyright@packt.com` with a link to the material.

If you are interested in becoming an author: If there is a topic that you have expertise in and you are interested in either writing or contributing to a book, please visit `authors.packtpub.com`.

Share Your Thoughts

Once you've read *Data Wrangling with SQL*, we'd love to hear your thoughts! Scan the QR code below to go straight to the Amazon review page for this book and share your feedback.

`https://packt.link/r/1-837-63002-X`

Your review is important to us and the tech community and will help us make sure we're delivering excellent quality content.

Download a free PDF copy of this book

Thanks for purchasing this book!

Do you like to read on the go but are unable to carry your print books everywhere? Is your eBook purchase not compatible with the device of your choice?

Don't worry, now with every Packt book you get a DRM-free PDF version of that book at no cost.

Read anywhere, any place, on any device. Search, copy, and paste code from your favorite technical books directly into your application.

The perks don't stop there, you can get exclusive access to discounts, newsletters, and great free content in your inbox daily

Follow these simple steps to get the benefits:

1. Scan the QR code or visit the link below

https://packt.link/free-ebook/9781837630028

2. Submit your proof of purchase
3. That's it! We'll send your free PDF and other benefits to your email directly

Part 1:
Data Wrangling Introduction

This part includes the following chapters:

- *Chapter 1, Database Introduction*
- *Chapter 2, Data Profiling and Preparation before Data Wrangling*

Database Introduction

Welcome to the exciting world of data-driven decision-making! In this fast-paced landscape, the ability to extract, transform, and analyze data efficiently is essential. At the heart of this process lies something incredibly powerful: databases. These structured repositories are key to organizing and managing vast amounts of information. If you want to make the most of your data-wrangling endeavors, understanding databases and the **Structured Query Language** (**SQL**) that brings them to life is crucial. That's where this book, *Data Wrangling with SQL*, comes in. It's a comprehensive guide, designed to empower you with the knowledge and tools you need to unlock the full potential of databases. By diving into the fundamentals of databases and SQL, you'll gain a deep appreciation for how crucial they are in the data-wrangling journey.

Getting started

Before delving into the fascinating world of data wrangling using SQL, it is essential to grasp the fundamental concepts of databases. This introductory chapter serves as the foundation for your data-wrangling journey, setting the stage for understanding why databases play a pivotal role in efficiently extracting insights from data.

Establishing the foundation

The study of databases forms the foundation upon which the entire data-wrangling process is built. Understanding the core concepts and principles of databases will enable you to navigate the intricacies of data management effectively. By familiarizing yourself with key terms such as tables, rows, and columns, you'll develop a solid foundation upon which you can confidently build your data-wrangling skills.

Efficient data organization

Databases provide a structured and organized approach to storing and retrieving data. They offer a systematic way to manage vast amounts of information, making it easier to store, update, and retrieve data when needed. By learning about database design principles, data modeling techniques, and normalization, you will be equipped with the knowledge to create efficient and optimized database structures, ensuring smooth data-wrangling processes.

Data integrity and consistency

In the field of data wrangling, maintaining data integrity and consistency is of utmost importance. Databases provide various mechanisms, such as constraints and relationships, to enforce data integrity rules and ensure the accuracy and reliability of the data. Having a clear understanding of how databases maintain data consistency can help you trust the quality of the data you are working with, leading to more reliable and meaningful insights during the data-wrangling process.

By understanding the essential concepts discussed in this introductory chapter, you will be well equipped to begin your data-wrangling journey using SQL. A solid understanding of databases will give you the confidence to tackle real-world data problems, ensuring that your data-wrangling efforts are accurate, reliable, and efficient. Therefore, let's delve into the material and establish a solid foundation for a productive and satisfying data-wrangling experience!

Technical requirements

Please follow these step-by-step instructions to install MySQL on your machine and create a database on it.

To install MySQL Workbench on your computer, follow these steps:

1. **Visit the official MySQL website**: Go to the MySQL website at `https://dev.mysql.com/downloads/workbench/`.

2. **Select your operating system**: On the MySQL Workbench downloads page, you will see a list of available operating systems. Choose the appropriate option for your operating system. For example, if you are using Windows, click on the **Windows (x86, 64-bit), MSI Installer** option.

3. **Download the installer**: Click on the **Download** button next to the selected operating system version. This will start the download of the MySQL Workbench installer file.

4. **Run the installer**: Once the download is complete, locate the downloaded installer file on your computer and double-click on it to run the installer.

5. **Choose installation options**: The installer will guide you through the installation process. You can choose the installation type (**Typical**, **Complete**, or **Custom**) and specify the installation location if desired. It is recommended to choose the **Typical** installation type for most users.

6. **Accept the license agreement**: Read through the license agreement and click on the checkbox to accept it. Then, click on the **Next** button to proceed.

7. **Install MySQL Workbench**: Click on the **Install** button to start the installation process. The installer will copy the necessary files and install MySQL Workbench on your computer.

8. **Complete the installation**: Once the installation is finished, you will see an **Installation Complete** message. Click on the **Next** button to proceed.

9. **Launch MySQL Workbench**: By default, the installer will offer to launch MySQL Workbench immediately after installation. If the option is selected, MySQL Workbench will open automatically. Otherwise, you can manually launch it from the **Start** menu or desktop shortcut.

That's it! MySQL Workbench is now installed on your computer. You can now launch it and start using it to connect to MySQL servers, manage databases, and perform various database-related tasks.

> **Note**
>
> For a step-by-step pictorial representation of setting up MySQL Workbench, please follow this link: `https://www.dataquest.io/blog/install-mysql-windows/`.

Decoding database structures – relational and non-relational

Before we delve into the details of relational and non-relational databases, let us first understand the meaning of the term *database* and why it is important to know about databases.

What is a database?

Most of us have heard of a database, right? To put it simply, it is a collection of information that is stored in an organized and logical manner. This helps people keep track of things and find information quickly. For example, imagine you are walking into a superstore and looking for a specific item, such as a phone charger. To find it, you would use logical categorization. First, you would go to the electronics section, but this section would have all sorts of electronics the superstore had to offer. So, you would then look for a section called phones and accessories and search for the specific phone charger that was compatible with your phone.

By using logical reasoning, you can determine the location of the object and purchase the charger successfully. If we consider the process from the perspective of the superstore, we can see that they have divided the entire area into sections such as electronics, merchandise, and groceries, and further subdivided it into rows and columns known as aisles. They store each object according to its category in an organized manner, which can be accessed through the store's database.

The business definition of a database is that it is a collection of information stored on a server that is accessed regularly for analysis and decision-making. The information is organized into tables, which are similar to spreadsheets, with rows and columns. A database can contain multiple tables, and a server can have multiple databases for different categories or clients. For example, a university database may contain information on students, teachers, and subjects, while a superstore database may contain data on products, orders, store locations, and customers. Each row in the database represents a specific occurrence or transaction. The database stores information and its relationships.

Types of databases

Database Management Systems (**DBMSs**) are used to store and manage data in a database. The most commonly used language to extract information from a database is SQL. The history of databases dates back several decades, specifically to the 1970s. Since then, databases have evolved into two broad categories, known as relational and non-relational.

Relational databases

A relational database, or relational DBMS, stores data in the form of tables or entities that we want to track, such as customers and orders. The data about these entities is stored in relations, which are 2D tables of rows and columns, similar to a spreadsheet. Each row contains data, and each column contains different attributes about that entity.

SQL DATABASES

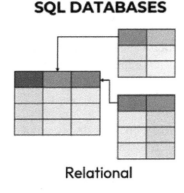

Relational

Figure 1.1 – Relational database

For instance, in a table/entity that contains information about customers, the attributes or columns could include `Name`, `Phone Number`, `Address`, and `Gender`. The rows would then represent specific information for each customer in a separate row.

For example, we could have a customers table as follows:

Customer_ID	Name	Address	Phone	Gender	Email
1	Joey	Texas	834-2345	M	JT@domain.com
2	Ron	Tennessee	987-6543	M	RT@domain.com
3	Fred	New York	876-5678	M	FN@Domain.com
4	Tom	LA	765-7654	M	TL@domain.com
5	Mary	Georgia	124-0987	F	MG@domain.com

Figure 1.2 – Customers table

Every row in a relational database should have a unique key, which we call the primary key (discussed later in the chapter). This key can be used as a foreign key in a different table to build logical referential relations between the two tables. The relations between the fields and tables are known as schemas. To extract data from databases, we use SQL queries.

These are some of the advantages of relational databases:

- Highly efficient

- High readability as data is sorted and unique

- High data integrity

- Normalized data

Non-relational databases

A non-relational database stores data in a non-tabular format, meaning it does not have a structure of tables and relations. Instead, this type of database stores information in various ways, such as key-value and document-based databases. In a key-value database, data is stored in two parts: a key and its corresponding value. Each key is unique and can only connect to one value in the collection. In contrast, a document-oriented database pairs a key with a document that contains a complex combination of several key-value pairs. Non-relational databases, also known as NoSQL databases, are more flexible than traditional relational databases. Some commonly used non-relational databases include MongoDB, Cassandra, Amazon DynamoDB, and Apache HBase.

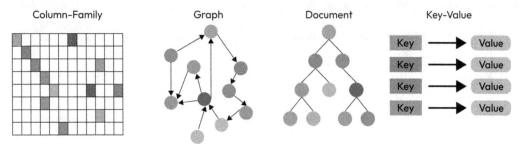

Figure 1.3 – NoSQL databases

Key	Document
| 2022 | ```
{
    Customer ID: 1234,
    Customer Name: Joey,
    Customer address: XYZTX,
    Order details:
      {
        Order 1: {product 1, product description}
        Order 2: {Product 1, product description}
      }
}
``` |
| 2023 | ```
{
    Customer ID:5667,
    Customer Name: Ron,
    Customer address: LKJHNTN,
    Order details:
      {
        Order 1: {product 1, product description}
        Order 2: {Product 1, product description}
      }
}
``` |

Figure 1.4 – Non-relational database example

These are some of the advantages of non-relational databases:

- Simple management – no sorting needed, so data can be directly dumped into the database without any preprocessing

- Higher readability of a particular document, especially when the dataset contains big data, avoiding the need to parse through multiple tables and write complex queries

- Can be scaled to a huge level by splitting the servers into multiple clusters and managing the CPU utilization on each of these clusters

Let's understand that last point in detail. *Multiple clusters* refers to a distributed computing architecture that consists of multiple servers or nodes, each serving a different purpose, but working together to achieve a common goal. In this context, a cluster typically consists of a group of interconnected computers that works together to provide a more powerful and scalable computing environment. Each cluster may have its own dedicated resources, such as CPU, memory, and storage, and can be managed independently. By splitting servers into multiple clusters, the workload can be distributed more efficiently, allowing for greater scalability and flexibility. For example, suppose you have a large-scale application that requires a lot of processing power and storage. In that case, you might split your servers into multiple clusters and distribute the workload across those clusters. This way, you can achieve better performance, as well as reduce the risk of a single point of failure. Overall, multiple clusters offer several advantages, including increased scalability, improved performance, better fault tolerance, and the ability to handle large workloads more efficiently.

Tables and relationships

In the realm of database management, tables and relationships form the backbone of organizing and connecting data in SQL Server. With the power of structured data organization and connectivity, SQL Server enables businesses and organizations to efficiently store, retrieve, and analyze vast amounts of information.

The SQL CREATE DATABASE statement

All tables in SQL Server are stored in a repository known as a database. A database is a collection of tables related to a specific entity. For instance, we can have separate databases for insurance company providers such as Blue Cross Blue Shield and Cigna. Having one database for each entity helps maintain and scale the database and its dataset for the future. The owner of the database is known as a database administrator who holds admin privileges to the database. Only a database administrator can provide or revoke access to a database.

- The syntax for creating a database is `CREATE DATABASE Database_name`
- The preceding statement will create a database with the name `Database_name`
- The syntax for deleting a database is `DROP DATABASE Database_name`
- The syntax for viewing the entire database is `SHOW DATABASE`

The SQL CREATE TABLE statement

The `CREATE TABLE` statement is used to create a new table, which is a combination of rows and columns, in a specified database:

```
CREATE TABLE DATABASENAME.TABLE_NAME
(
Column 1 datatype,
Column 2 datatype,
Column 3 datatype,
);
```

For example. we can create a `walmart.customer_info` table as follows:

```
(
Customer_ID int,
Name varchar(255),
Address varchar(255),
```

```
Email varchar(255),
Phone varchar(255)
)
```

In the preceding code, the `Customer_ID` column is defined as an integer data type, while the `Name` column is defined as a varchar data type. This means that the `Name` column can hold both letters and numbers, up to a maximum length of 255 characters. The code will create an empty table with these columns.

Customer_ID	Name	Address	Phone	Email

Figure 1.5 – customer_info table

> **Note**
>
> Post-creation of this table, the database administrator has to provide read access to all the table users so that they can access the data within it.

SQL DROP TABLE versus TRUNCATE TABLE

If the requirement is to delete the entire table along with its schema, the SQL syntax used is DROP TABLE `table_name`. However, if the need is to delete only the content from within the table and retain the table structure, the SQL syntax used is TRUNCATE TABLE `table_name`. After truncation, the table will still exist but with 0 rows within it.

SQL ALTER TABLE

The SQL syntax of ALTER TABLE is used to insert, update, or delete data or columns in a pre-created table:

```
ALTER TABLE Walmart.customer_info
DROP COLUMN Email;
```

The following table will be created after the execution of the drop statement:

Customer_ID	Name	Address	Phone

Figure 1.6 – Email column dropped

SQL constraints

Constraints are a set of predefined rules that a database administrator or table creator defines to ensure that the table and its data are unique and clean. They can be defined at the table or column level. Some commonly used constraints include the following:

- The **UNIQUE constraint**: Ensures all columns have a unique value.

- The **NOT NULL constraint**: Ensures all columns have some value.

- **PRIMARY KEY**: A unique value at each row level.

- **FOREIGN KEY**: A relational key, which is a copy of a primary key column from a different table within the same database. A foreign key is used to ensure that the communication between two tables is maintained and never destroyed by accidentally dropping the foreign key column. More on this will be discussed in the following sections.

SQL keys

In RDBMSes, the term "relational" refers to the relationship between tables in a database that allows for the retrieval of necessary data. This relationship is established through the use of keys such as primary keys, foreign keys, and candidate keys.

Customer_Id	Name	Passport_Number	DOB
1	Adam	L08790	7/11/1990
2	James	L08791	8/6/1992
3	Paul	L08792	3/4/1993

Figure 1.7 – Candidate keys

Candidate keys

A candidate key is a set of one or more columns that can uniquely identify a record in a table. It can be used as a primary key as it cannot be null and must be unique. A candidate key is a super key with no repeating attributes. Out of all the candidate keys that are possible for a table, only one key can be used to retrieve unique rows from the table. This key is called the primary key. It is important to remember that the candidate key used as a primary key should be unique and have non-null attributes.

In the customer table shown in *Figure 1.8*, we have *one candidate key* – **Passport_Number** – which is *unique*, whereas the **Customer_ID** column is a *primary key*.

Customer_ID	1	2	3
Name	Adam	James	Paul
Passport_Number	L08790	L08791	L08792
DOB	7/11/1990	8/6/1992	3/4/1993

Figure 1.8 – Candidate key

Primary keys

A primary key is an attribute used to uniquely identify a row in a table. In the table mentioned previously, **Customer_ID** would be the primary key, while **Passport_Number** would not be a primary key as it contains confidential information.

Customer_ID	1	2	3
Name	Adam	James	Paul
Passport_Number	L08790	L08791	L08792
DOB	7/11/1990	8/6/1992	3/4/1993

➡ **Primary Key**

Figure 1.9 – Primary key

Alternate keys

A candidate key that hasn't already been assigned as a primary key is known as an alternate key that can uniquely identify a row. In the following table, **Customer_ID** is the primary key and **Passport_Number** is a candidate key; therefore, **License_Number** can be an alternate key as it can also uniquely identify a customer.

Customer_ID	1	2	3
Name	Adam	James	Paul
Passport_Number	L08790	L08791	L08792
DOB	7/11/1990	8/6/1992	3/4/1993
License_Number	L01YZ	L02ER	L03PX

➡ **Alternate Key**

Figure 1.10 – Alternate key

Super keys

If more than one attribute is assigned as the primary key and it still uniquely identifies a row within a table, then it becomes a super key.

For example, `Customer_ID + Name` is a super key, as the name of a customer may not be unique, but when combined with `Customer_ID`, then it becomes unique.

Customer_ID	1	2	3
Name	Adam	James	Paul
Passport_Number	L08790	L08791	L08792
DOB	7/11/1990	8/6/1992	3/4/1993

} ➡ **Super Key**

Figure 1.11 – Super key

Composite keys

If the table does not have an individual attribute that can qualify as a candidate key, then we need to select two or more columns to create a unique key, which is known as a composite key.

For example, if we do not have a customer ID or passport number, we can use a composite primary key consisting of the full name and date of birth.

There is still the possibility of duplicate rows in this scenario if both the name and date of birth have the same value.

Customer_ID	1	2	3
Name	Adam	James	Paul
Passport_Number	L08790	L08791	L08792
DOB	7/11/1990	8/6/1992	3/4/1993

Figure 1.12 – Composite key

Surrogate key

A surrogate key is a key that is generated by the system and has no business meaning. The values generated for the keys are sequential and act as a primary key. When we don't have a proper primary key for the table, a surrogate key is generated to uniquely identify the data. In such scenarios, the surrogate key becomes the primary key.

For example, let's consider that we are creating a database for addresses.

Address_ID	Street_Name	City	State	Zipcode
1	Jefferson St	Dallas	Texas	38256
2	Thomas St	Memphis	Tennessee	38257
3	James St	Chicago	Illinois	33189
4	Perkins St	Miami	Florida	23487

Figure 1.13 – Address table with Address_ID as a surrogate key

Here Address_ID is the surrogate key as it is generated systematically and is used to uniquely identify the rows. It holds no business value.

Primary keys in detail

A primary key is a specially created attribute to uniquely identify each record in a table and has the following features:

- It holds unique values for each record/row

- It can't have null values

The primary key and foreign key are core principles used to establish relationships between tables in a relational database.

A few examples of primary keys include **Social Security Number** (**SSN**), passport number, and driver's license number. These are used to uniquely identify a person.

To ensure unique identification, a composite primary key is sometimes created using a combination of columns.

Foreign keys in detail

A foreign key is a column or combination of columns that creates a link between data in one table (the referencing table) and another table that holds the primary key values (the referenced table). It creates cross-references between tables by referencing the primary key (unique values) in another table.

The table that has the primary key is called the parent table or referenced table, and the table that has a foreign key is called the referencing table or child table.

The column that has a foreign key must have a corresponding value in its related table. This ensures referential integrity.

The FOREIGN KEY constraint is used to prevent actions that would destroy links between tables. Essentially, it ensures that if a column value A refers to a column value B, then column value B must exist.

For example, let's consider the Orders table and the Customers table. In this case, the customer_ID column in the Orders table refers to the Customer_ID column in the Customers table. Here are some key points regarding the relationship between these tables:

- Any value that is updated or inserted in the customer_id attribute (foreign key) of the Orders table must exactly match a value in the Customer_ID attribute (primary key) of the Customers table or be NULL to ensure the relation is maintained.

- The values in the ID attribute (the Customer_ID primary key) of the Customers table that are referencing the customer_ID attribute (foreign key) of the Orders table cannot be updated or deleted unless cascading rules are applied (cascading actions will be discussed shortly). However, the values of ID in the Customers table that are not present in the Customer_ID attribute of the Orders table can be deleted or updated.

Primary_Key		
Customer_ID	**Name**	**Address**
1	Joey	Texas
2	Ron	Tennessee
3	Fred	New York
4	Tom	LA
5	Mary	Georgia

Foreign_key			
Order ID	**Customer_ID**	**Order Date**	**Shipping Status**
01	1	1/1/2022	Delivered
02	1	9/1/2022	In Progress
03	2	12/20/2022	Not Started
04	3	8/15/2022	Delivered
05	4	5/31/2022	Delivered

Figure 1.14 – Foreign key illustration

Cascading actions

Cascading actions in SQL refer to the automatic propagation of changes in a parent table to related child tables through foreign key constraints. It enables actions such as deletion or modification in the parent table to automatically affect corresponding records in the child tables. This ensures data integrity and simplifies data management by reducing manual updates.

DELETE CASCADE

This ensures that when a row with a primary key is deleted from a parent table, the corresponding row in the child table is also deleted.

UPDATE CASCADE

This ensures that when a referencing row that is a primary key is updated in a parent table, then the same is updated in the child table as well.

Using a **foreign key** eliminates the need to store data repeatedly. Since we can directly reference primary keys in another table, we don't have to store that data again in every table.

Take the following example:

Customer_Id	Name	Address	Phone	Gender	Email
1	Joey	Texas	834-2345	M	JT@domain.com
2	Ron	Tennessee	987-6543	M	RT@domain.com
3	Fred	New York	876-5678	M	FN@Domain.com
4	Tom	LA	765-7654	M	TL@domain.com
5	Mary	Georgia	124-0987	F	MG@domain.com

Figure 1.15 – Customers table

Order_ID	Customer_ID	OrderDate	ShippingDate	ShippingStatus
O1	1	1/1/2022	1/7/2022	Delivered
O2	1	9/1/2022	9/4/2022	In Progress
O3	2	12/20/2022	12/31/2022	Not Started
O4	3	8/15/2022	8/20/2022	Delivered
O5	4	5/31/2022	5/31/2022	Delivered

Figure 1.16 – Orders table

As you can see, `Customer_ID` in the `Orders` table is a foreign key that can be used to establish a connection between the `Orders` and `Customers` tables. This allows us to retrieve customer details from the `Orders` table and vice versa.

Database relationships

Database relationships are used to build well-defined table structures and establish relationships between different tables. With the correct relationships, it helps to standardize data quality and eliminate data redundancy.

Different types of database relationships include the following:

- One-to-one relationships
- One-to-many relationships
- Many-to-many relationships

A database entity can be a customer, product, order, unit, object, or any other item that has data stored in the database. Typically, entities are represented by tables in the database.

An **entity relationship diagram**, commonly known as an **ER diagram** or **ERD**, is a flowchart that illustrates how different entities/tables are related to each other.

One-to-many relationships

One-to-many is the most common type of relationship, in which one record in one entity/table can be associated with multiple records in another entity/table.

Figure 1.17 – One-to-many relationship

Example

Let's consider the Customers and Orders tables. In this case, one customer can have multiple orders, establishing a one-to-many relationship between them. Customer ID in the Customers table serves as the primary key and is associated with unique values, representing each customer uniquely. On the other hand, Customer ID in the Orders table acts as the foreign key and can have multiple instances, indicating that multiple orders can be associated with the same customer.

In this case, a single customer ordered multiple products, creating a one-to-many relationship where each product was associated with one customer.

One-to-one relationships

A one-to-one relationship is a type of relationship between two tables in which one record in one table is associated with only one record in another table.

Figure 1.18 – One-to-one relationship

Example

In a school database, each student is assigned a unique student_ID, and each student_ID is linked to only one student.

In a country database, each country is associated with one capital city, and each capital city is associated with only one country.

Many-to-many relationships

A many-to-many relationship is one in which multiple records in one table are associated with multiple records in another table.

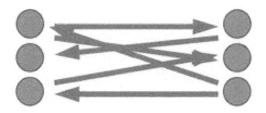

Figure 1.19 – Many-to-many relationship

Example

Let's consider the Customers and Products tables. Customers can purchase multiple products, and each product can be purchased by different customers.

In a relational database, a direct many-to-many relationship is typically not permitted between two tables. For example, in a bank transaction database with multiple invoices having the same number, it can be difficult to map the correct invoice and retrieve the necessary information when a customer makes an inquiry. To address this issue, many-to-many relation tables are often broken down into two one-to-one relationship tables by introducing a third table known as a join table. The join table holds the primary key of both tables as a foreign key and may also contain other necessary attributes.

For example, let's consider the Products and Distributor tables.

Here we have the following attributes in each table.

The attributes for the Distributor table are as follows:

- id: The distributor's ID is the primary key used to identify the distributor
- distributors_name: The distributor's name
- distributors_address: The distributor's address
- distributors_city: The city where the distributor is located
- distributors_state: The state where the distributor is located

The attributes for the Products table are as follows:

- id: The product's ID is the primary key used to identify the product ID
- product_name: The product's name
- product_description: The product's description
- price: The product's price per unit

Multiple products can be ordered by multiple distributors, and each distributor can order different products. Therefore, this information needs to be transformed into a relational database model, which would look something like this:

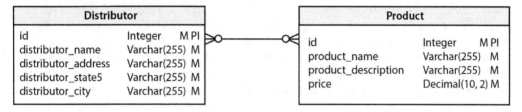

Figure 1.20 – Relational database model

This is known as a join table, and it contains two attributes that serve as foreign keys referencing the primary keys of the original tables.

id references the distributor ID from the distributor table.

product_id references the product_id column in the product table.

These two together serve as the primary key for this table.

However, this information is not sufficient. It would be better to add more attributes to this table.

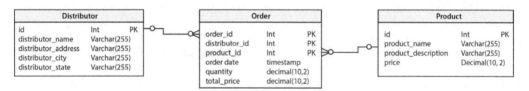

Figure 1.21 – Adding more attributes to the relational database model

So, we have now converted it into a relational database model and have the data in a cleaner form. Additionally, I have changed the table name to make it more aligned with the data and named it Orders. We have also added the following additional attributes to the table:

- order_id: Contains the unique order ID for each order placed by the customer
- distributor_id: The distributor's ID is the unique identifier for each distributor
- product_id: The unique identifier for each product, which can be ordered
- order_date: The order date
- quantity: The number of units ordered
- total_price: The total price of the orders

> **Note**
>
> There are two ways that you can create a join table.
>
> The first way to create a join table is by having the table contain only foreign keys that reference other tables. This is the most common and straightforward approach. In this case, the join table primarily serves as a mapping table, capturing the relationships between the records in the two tables it connects. For example, let's consider two tables: "Students" and "Courses." If multiple students can enroll in multiple courses, we need a way to represent this many-to-many relationship. We can create a join table named "Enrollments" that contains foreign keys referencing the primary keys of both the "Students" and "Courses" tables. Each row in the "Enrollments" table represents a specific student's enrollment in a particular course. The second way to create a join table involves adding additional attributes, effectively turning the join table into a new entity. In this approach, the join table gains its own meaning and significance beyond simply representing the relationship between the two tables. For instance, let's consider a scenario where we have two tables: "Employees" and "Projects." Instead of just recording the relationships between employees and projects, we may want to include additional attributes in the join table, such as the date an employee was assigned to a project or their role on that project. In this case, the join table, which we can call "Assignments," becomes a separate entity that stores specific information about the association between employees and projects. By adding these additional attributes, the join table becomes more than just a mapping table and can provide more context or details about the relationship between the two tables it connects.
>
> You should always model tables according to the requirements, and there are no hard and fast rules for doing so.

Comparing database normalization and denormalization

Lets now deep dive and understand the fundamental difference between normalization and de-normalization.

Normalization

Database design is a process that aims to reduce data redundancy and maintain data integrity. This systematic approach involves removing data redundancy and undesirable characteristics, such as data insertion, update, and delete anomalies. It achieves this by breaking down larger tables into smaller ones and linking them based on relationships to store data logically.

Data normalization is necessary to eliminate the following anomalies.

Insertion anomalies

An insertion anomaly occurs in relational databases when we are unable to insert data into the database due to missing attributes or data. This is a common scenario where a foreign key cannot be NULL but does not have the necessary data.

For instance, suppose a customer has a customer ID as a foreign key in the `Orders` table, and no customers have been inserted yet because they haven't ordered any products yet. Therefore, it is impossible to insert a customer who hasn't ordered anything yet.

Update anomalies

An update anomaly occurs when we partially update data in a database. This is a common scenario where data is not normalized, and hence, data elements can reference the same data element in more than one place. As all these different locations are not updated automatically, it is important to manually update this data element at each location.

This process can be time-consuming and inefficient as it requires searching for data elements in various locations and manually updating each one.

For instance, suppose we have a `Customers` table with five columns, including `Customer Phone Number` and `Customer Address`. If a customer's address or phone number changes, we must update the table. However, if the table is not normalized, a single customer may have multiple entries, and updating all of them could result in an update anomaly if one of them is overlooked.

Delete anomalies

Data loss can occur when important information is accidentally deleted along with other data. This can result in the loss of crucial data.

For example, let's consider a customer named Adam who ordered one product, a phone. If the order is canceled and you delete the customer from the order table, it will also delete the product (the phone).

Hence, to avoid these anomalies, we need to normalize the data.

Types of normalization

There are four types of normalization:

- **First Normal Form (1NF)**
- **Second Normal Form (2NF)**
- **Third Normal Form (3NF)**
- **Boyce–Codd Normal Form (BCNF)**

Let's explore each of them.

1NF

In 1NF, each attribute in the table should contain only atomic values. This means that each cell in the table should hold only one value, and the intersection of rows and columns should not contain multiple values or repeating groups.

For example, in the following table, we have details of the products purchased by customers.

Customer	Customer_Name	Products
1	Adam	Phone, Pen
2	James	Car, Ipod
3	Paul	Laptop, Cup

Figure 1.22 – Non-normalized product table

We can see that the `Products` attribute holds information related to multiple products purchased by the customer and is therefore not normalized. This is because it contains more than one value in a single cell.

Hence, this can be normalized in the following way in 1NF.

Customer ID	Customer_Name	Products
1	Adam	Phone
1	Adam	Pen
2	James	Car
2	James	Ipod
3	Paul	Laptop
3	Paul	Cup

Figure 1.23 – 1NF normalized product table

As we can see, each cell now holds only one value at the intersection of a column and row. Therefore, the data has been normalized to 1NF.

2NF

There are two rules that must be followed to normalize a table into 2NF:

- Rule 1 – the table first has to be in 1NF.

- Rule 2 – the table should not have any partial dependency. This means that every non-prime attribute (all attributes other than the primary key) must be dependent on the primary key of the table.

Whenever the table represents data for two different entities instead of just one entity, it needs to be broken down into its own entity in a different table.

For example, the following table has composite primary keys: **Customer_ID** and **Order_ID**. However, the non-prime attributes are not solely dependent on the **Customer_ID** primary key, but also on **Order_ID**. For instance, the **Order_Status** attribute, which is a non-prime attribute, is only dependent on **Order_ID**. Therefore, it is necessary to split the table into two separate tables. One table will contain the customer details, while the other will contain the order details.

Customer_ID	Customer_Name	Customer_Phone_Number	Order_ID	Order_Status
1	Adam	485-000-9890	1	In Progress
1	Adam	585-000-9890	2	Delivered
2	James	685-000-9890	3	In Progress
2	James	785-000-9890	4	In Progress
3	Paul	885-000-9890	5	Delivered
3	Paul	985-000-9890	6	Delivered

Figure 1.24 – Customer details table

The table depicted in *Figure 1.24* has been decomposed into two separate tables that satisfy 2NF.

Customer_ID	Customer_Name	Customer_Phone_Number
1	Adam	485-000-9890
1	Adam	585-000-9890
2	James	685-000-9890
2	James	785-000-9890
3	Paul	885-000-9890
3	Paul	985-000-9890

Figure 1.25 – Customers table, which holds customer details

Order_ID	Order_Status
1	In Progress
2	Delivered
3	In Progress
4	In Progress
5	Delivered
6	Delivered

Figure 1.26 – Orders table

3NF

3NF ensures a reduction in data duplication, thereby maintaining data integrity by following these rules:

- Rule 1 – the table has to be in 1NF and 2NF.

- Rule 2 – the table should not have transitive functional dependencies. This means that non-prime attributes, which are attributes that are not part of the candidate key, should not be dependent on other non-prime attributes within the table.

In the following table, `Customer_ID` determines `Product_ID`, and `Product_ID` determines Product_Name. Hence, `Customer_ID` determines `Product_Name` through `Product_ID`. This means the table has a transitive dependency and is not in 3NF.

Therefore, the table has been divided into two separate tables to achieve 3NF.

Customer_ID	Customer_Name	Product_ID	Product_Name	Customer_Phone_Number
1	Adam	1	Phone	485-000-9890
1	Adam	2	Pen	585-000-9890
2	James	3	Car	685-000-9890
2	James	4	iPod	785-000-9890
3	Paul	5	Laptop	885-000-9890
3	Paul	6	Cup	985-000-9890

Figure 1.27 – Customer order table

Customer_ID	Customer_Name	Customer_Phone_Number
1	Adam	485-000-9890
1	Adam	585-000-9890
2	James	685-000-9890
2	James	785-000-9890
3	Paul	885-000-9890
3	Paul	985-000-9890

Figure 1.28 – Customers table

Product_ID	Product_Name
1	Phone
2	Pen
3	Car
4	iPod
5	Laptop
6	Cup

Figure 1.29 – Products table

Therefore, it is evident that non-prime attributes, which refer to attributes other than the primary key, are solely reliant on the primary key of the table and not on any other column. For instance, non-key attributes such as `Customer_Name` and `Customer_Phone_Number` are solely dependent on `Customer_ID`, which serves as the primary key for the `Customers` table. Similarly, non-key attributes such as `Product_Name` are exclusively dependent on `Product_ID`, which acts as the primary key for the `Products` table.

BCNF

BCNF is sometimes referred to as 3.5 NF. Even if a table is in 3NF, there may still be anomalies present if there is more than one candidate key.

Two rules are to be followed for the table to be in BCNF:

- Rule 1 – the table should be in 1NF, 2NF, and 3NF
- Rule 2 – for every functional dependency, such as A -> B, A should be the super key

Super key

Consider the following table:

Student_ID	Subject	Faculty
Student_1	Cloud Computing	Professor A
Student_2	Big Data	Professor B
Student_3	Statistics	Professor C
Student_4	Project Management	Professor D
Student_5	Analytics	Professor E

Figure 1.30 – Student-subject table

In the table depicted in *Figure 1.30*, a student can study multiple subjects in one semester, and multiple professors can teach one subject. For each subject, one professor is assigned. All normal forms are satisfied except for BCNF. The primary key is formed by combining `Student_ID` and `Subject`, as this allows for the unique extraction of information related to both faculty and students. Additionally, there is a dependency between the `Subject` and `Faculty` columns. This is because one subject can be taught by multiple professors, creating a dependency between the two columns.

There are a few key points to understand here:

- The preceding table is in 1NF as all rows are atomic
- The preceding table is in 2NF as there is no partial dependency
- The preceding table is also in 3NF as there is no transitive dependency

This table does not satisfy the BCNF condition because there is a dependency of Faculty on the Subject, which makes the Faculty column a non-prime attribute while Subject is a prime attribute.

How to fix this

A new super key called `Faculty_ID` will be introduced, and the preceding table will be split into two different tables, as follows.

Student_ID	Faculty_ID
Student_1	Faculty_ID1
Student_2	Faculty_ID2
Student_3	Faculty_ID3
Student_4	Faculty_ID4
Student_5	Faculty_ID5

Figure 1.31 – Student-faculty mapping table

Faculty_ID	Faculty	Subject
Faculty_ID1	Professor A	Cloud Computing
Faculty_ID2	Professor B	Big Data
Faculty_ID3	Professor C	Statistics
Faculty_ID4	Professor D	Project Management
Faculty_ID5	Professor E	Analytics

Figure 1.32 – Faculty table

Now, the `Faculty_ID` super key has eliminated the non-prime attribute functional dependency and the BCNF normal form is satisfied.

Denormalization

Before delving into denormalization, let's first quickly review normalization, which we just discussed. Normalization, in simple terms, involves segmenting each piece of data into the appropriate bucket, which helps maintain data integrity, eliminates redundancy, and ensures easy updates to the data in the future.

However, the drawback of using normalized data is that it may take a longer time to retrieve the data as the data has to be queried from different tables where it is stored.

Denormalization comes into play when there is a trade-off between organized data and faster data retrieval. It is the opposite of normalization. In simple terms, denormalization involves storing the same data in multiple locations, which increases redundancy but decreases data retrieval time, making it faster.

Denormalization is used when there are performance issues with normalized data or when faster data retrieval is needed.

For example, let us consider the "`Customers`" and "`Orders`" tables. Both of these are currently normalized as there is no redundancy, and the data is housed in only one table, maintaining data integrity between the primary key and foreign key.

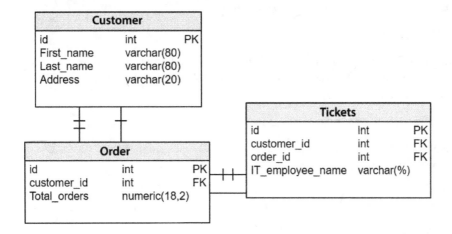

Figure 1.33 – Normalized tables

In the following table, we have two new tables: the customer census table and the IT incidents table. Let's take a closer look at the type of data that each of these tables contains:

- **customer_census**: The table contains statistics related to customers, including the number of orders placed, the total amount spent on orders, the number of incidents raised due to complaints, and the total number of complaint calls made by the customer. This table provides information on customer orders and complaint statistics.

- **IT incidents**: The table holds information related to the tickets raised by all customers for different orders, customers, and the customer service agent working on the ticket.

Both of these tables are denormalized, as we have included both customer and order information. There is no need to join these tables with the "Customer" or "Order" tables, as they contain all the necessary information. Retrieving data from these tables is faster than from normalized tables.

> **Interesting read**
>
> These types of tables are utilized by e-commerce companies to monitor customer service metrics and enhance customer service on a **month-over-month** (**M-O-M**), **quarter-over-quarter** (**Q-O-Q**), and **year-over-year** (**Y-O-Y**) basis.

A few of the main **key performance indicators** (**KPIs**) to track in this area are as follows:

- Customer call frequency over a period of time
- Number of incidents raised over a period of time
- Major dispute reasons

These two tables contain redundant data, such as customer and order information repeated multiple times. However, this data can be retrieved from a single table, making the process faster. This is a clear example of denormalized tables.

Whenever order details are modified, the changes must be reflected in both of the new tables.

Order details need to be updated in new tables as well to maintain data integrity. This is one of the disadvantages of denormalized tables.

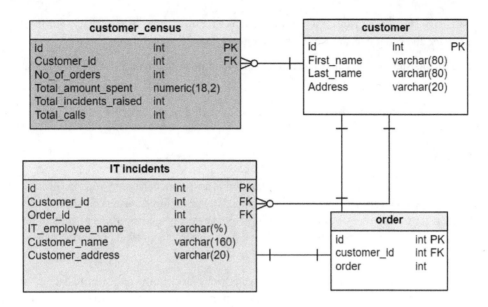

Table 1.34 – Denormalized tables

When to apply denormalization

Some situations in which it may be appropriate to apply denormalization are as follows:

- **Performance improvement**: One of the main reasons to use denormalization is to improve query performance. In denormalization, we read data from fewer tables as data is available in multiple tables. Hence, denormalized tables have better performance.

- **Maintaining history**: Maintaining history in different tables helps with data validation. For example, if we have built a dashboard using these tables, and we want to look back at the history and validate it in the future, denormalized tables can help with validation.

Disadvantages of denormalization

Here are some of the disadvantages of denormalization:

- **Expensive updates and inserts**: Whenever data is updated in one table, it needs to be updated in other related tables as well. For instance, if customer or order data is updated in one table, it must also be updated in all other tables where it is stored. This process can be costly and time-consuming.

- **Expensive storage**: Storing data in multiple tables requires more storage space. For instance, customer and order data are stored in separate tables.

- **Data inconsistency**: This can occur when data is not updated correctly, leading to discrepancies and errors in the data.

With that, we have reached the end of the chapter. Let's conclude everything we have learned so far.

Summary

In conclusion, this chapter covered several key foundational aspects of databases. We learned about the basics of databases, including the different types that exist. We explored various database attributes, such as fields, records, tables, databases, relationships, and key fields such as primary keys and foreign keys. Understanding these components is crucial for effectively structuring and organizing data within a database.

Additionally, we discussed the importance of following rules for defining and storing data in tables. Adhering to these rules ensures efficient data manipulation and analysis within the database. Proper data integrity, consistency, and table relationships enable seamless data wrangling and analysis processes.

By grasping these fundamental concepts, we have laid the groundwork for further exploration and understanding of databases and their role in managing and leveraging data effectively.

In the next chapter, we will learn about the prerequisites that should be completed before performing data wrangling. This step is known as data profiling, where we essentially understand and examine the data within the database.

Practical exercises

Before you start with the following exercises, please ensure that you have successfully completed the database setup mentioned in the *Technical requirements* section.

Practical exercise 1

Create a database with the name `Packt_database` and then verify whether the database is created correctly:

```
CREATE DATABASE Packt_database;
SHOW DATABASE;
```

Practical exercise 2

Create two tables within `Packt_database`, one customer and one product table, with the following columns and data types:

- Customer table: `Customer_ID`, `Customer_name`, `Customer_country`, and `Customer_gender`

- Product table: `Product_ID`, `product_description`, and `Customer_ID`

```
Create table packt_database.customer
(
Customer_ID int primary key identity (1,1),
Customer_name varchar (255),
Customer_country varchar (255),
Customer_gender varchar (2)
);
Create table packt_database.product
(
Product_ID int primary key identity (1,1),
Product_description varchar (255),
Foreign key (Customer_ID) REFERENCES customer(Customer_ID)
);
```

Practical exercise 3

Add an additional column named `product_cost` to the product table created in the previous exercise:

```
ALTER TABLE packt_database.product ADD product_cost varchar(255);
```

Practical exercise 4

You can continue practicing the learned concepts from SQL zoo link below:

```
https://sqlzoo.net/wiki/SQL_Tutorial
```

2

Data Profiling and Preparation before Data Wrangling

Now that we have a good idea about how data is stored and maintained in a database and how normalization and de-normalization are used to store data, in this chapter we will discuss the next stage in the process, which is cleaning and transforming data.

In this chapter, we will cover the following main topics:

- Data wrangling and its importance
- Structured and unstructured data
- Data wrangling tools that are used in the industry
- What is data profiling?

What is data wrangling?

Data wrangling is the process of cleaning, transforming, and organizing dirty data into clean data that can be used to generate powerful insights to enable stakeholders to make the right decisions. It is basically the process of removing errors in data and making it ready for analysis. As the amount of data is growing exponentially throughout the world, it is becoming more and more important to store and organize these large datasets properly. Real-world data is often quite messy and unstructured, hence it needs to be cleaned before it can be used for any analysis.

Figure 2.1 – Data wrangling

Let's look at a few examples of data wrangling:

- Cleaning dirty data, such as missing values, bad characters, unmatched data types, and bad formatting into consistent and clean data

- Combining different datasets from multiple sources and making sure data is consistent

- Deleting data that is no longer required

- Identifying outliers within data and cleaning data further

Example scenario

Let's say we have a hotel database and we need to analyze the data for a specific hotel, let's call it Xostl, where customers provide details such as phone numbers, names, and the dates of their stay. All this data can be entered in different formats. Some might enter data in uppercase, some in lowercase, and some of them might enter the date as 27-Jan-2018 whereas others might enter the same date as 27/01/2018. This is where data wrangling comes into play, which makes data more reliable and accurate for analysis.

Data wrangling steps

Several steps need to be followed in the data wrangling process so that data becomes useful for analysis.

Figure 2.2 – Data wrangling steps

In this section, we will be walking through the different steps of wrangling using the Customers table shown in *Figure 2.3*.

Discovery

This is the first step in data wrangling. During this process, we can find missing values, bad characters, unmatched data types, and different trends and patterns in the data. This is the most important step in data wrangling; it helps you to understand the data properly and see where it needs to be cleaned.

Example scenario

If you are looking at customer data, you need to understand what purchases the customer usually makes, what products the customer has bought, and everything related to the data. During the Discovery step, we learn the meaning of the data from a business point of view. In this case, in the table shown in *Figure 2.3*, we learn that it holds customer details. Then, we look for dirty data. As we can see, we have missing phone number values, bad characters in the Gender column, and one duplicate row (for **customer ID 3**). In this way, we can discover patterns within the data and understand the data thoroughly.

These are the different scenarios we can face in the real world with huge datasets. Hence, to make the data consistent, it is important to wrangle and clean the data into a consistent format.

Structuring

In this step, you transform your data from raw data into better formatted or cleaner data based on patterns observed in the Discovery step. Think of it as giving the right structure to the data so that it can be used by the stakeholders.

Example scenario

In this step, we could need to arrange customer data from different Excel files and organize that data into one consistent format.

Customer_ID	Name	Address	Phone	Gender	Email
1	Joey	Texas	902-834-2345	M	JT@domain.com
2	Ron	Tennessee	0	M	RT@domain.com
3	Fred	New York	902-876-5678	M	FN@Domain.com
4	Kim	LA	902-765-7654	F	TL@domain.com
5	Mary	Georgia	0	F	MG@domain.com

Figure 2.3 – Customers table

In the Customers table, we can clearly see that not all the column headings are clear, and they can be structured into a better format, as shown in *Figure 2.4*.

Cleaning

Data cleaning is the process of removing any errors in the data that might negatively impact our ability to draw meaningful inferences from the data. It is in this stage that we also deal with inconsistent data and try to convert it into a consistent format.

Example scenario

In the Cleaning stage, we may need to remove outliers, delete data such as missing values, standardize the data using imputation techniques, and various other tasks. This is an important step that minimizes errors within data.

The table in *Figure 2.3* has a lot of dirty data that needs to be put into the right format. In *Figure 2.4*, missing values have been replaced with null values. The /**m** in the Gender column is in a different format to the rest of the data in the Gender column. We are supposed to have either M or F in these columns, but here we have /**m**. Hence, we will convert this into M. Here, we are dealing with inconsistent data by converting it into a format consistent with the rest of the values in the column.

Customer_ID	Name	Address	Phone	Gender	Email
1	Joey	Texas	902-834-2345	M	JT@domain.com
2	Ron	Tennessee		M	RT@domain.com
3	Fred	New York	902-876-5678	M	FN@Domain.com
4	Kim	LA	902-765-7654	F	TL@domain.com
5	Mary	Georgia		F	MG@domain.com

Figure 2.4 – Customers table

Enriching

Once the data is structured and cleaned into a more usable format, you need to check whether you have all the data required for the project or whether you need to enrich data by extracting values from different datasets. This process is known as the enrichment of data. If you wish to continue with the Enriching step, you need to make sure the previous three steps are followed for the data extracted from different datasets.

Example scenario

Getting additional data, such as the shipping status of various orders of customers from a different database, can lead to better insights.

In a real-world situation, you may need to extract additional customer data from different datasets and follow the Discovery, Structuring, and Cleaning steps to make sure they are in the right format before combining the data.

Validating

Data validation is the process of ensuring data has the right quality and consistency. During this process, you need to make sure several different checks are done to ensure the highest quality of the data.

Figure 2.5 – Examples of dirty data

Several different checks can be done:

- Missing values
- Duplicates
- Bad characters
- Data types
- Formatting
- Logical conditions
- Data consistency

During this stage, it is possible that you might find a few errors that you will need to resolve. This process can be automated using a tool such as Alteryx. We will discuss this tool toward the end of this chapter.

Publishing

Once all the preceding steps have been followed and you are confident after the validations that the data is of the highest quality, you can go ahead and publish the data. Once you publish the data, it will be available for different people within the organization based on security access. The method of sharing the data will depend on your project goals and can be in the form of dashboards, reports, Excel files, etc., which would help stakeholders to make business decisions. After this step, the data is ready for analysis.

Customer_ID	Name	Address	Phone	Gender	Email
1	Joey	Texas	902-834-2345	M	JT@domain.com
2	Ron	Tennessee		M	RT@domain.com
3	Fred	New York	902-876-5678	M	FN@Domain.com
4	Kim	LA	902-765-7654	F	TL@domain.com
5	Mary	Georgia		F	MG@domain.com
6	Tim	Texas	902-834-2345	M	JT@domain.com
7	Lee	California	902-834-2346	M	RT@domain.com
8	July	Washington	902-876-5678	F	FN@domain.com
9	Julia	LA	902-765-7654	F	TL@domain.com
10	Sam	Georgia	902-723-7827	F	MG@domain.com

Figure 2.6– Customers table after data wrangling

The importance of data wrangling

Now, let's look at the diverse benefits of data wrangling.

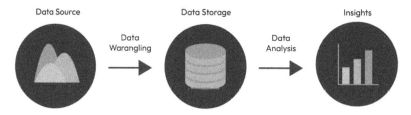

Figure 2.7 – Data wrangling process

> **Interesting read**
>
> Do you know that 80% of data analysts' time is spent on data wrangling, and just 20% of their time is spent on analysis? It is like building a house, where 80 % of the time is spent on constructing the main foundations and structure, and 20% of the time is spent on the final stages, such as painting. Similarly, as an analyst, most of your time needs to be spent on building the right foundation and structure; if the data is not usable for analysis, it can have a huge impact on the time and cost for the organization. Hence, data wrangling plays a key role in giving life to data.

Data quality

Maintaining data quality means ensuring the right data standards are met for analysis and the highest quality is maintained, reducing the errors in the data, thereby increasing the trustability and reliability of the data.

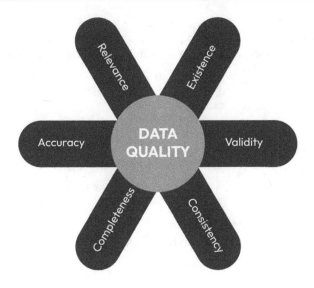

Figure 2.8 – Data quality

Multiple data sources

Getting data from multiple different sources into a centralized location requires proper data wrangling to ensure consistency is maintained while combining the data.

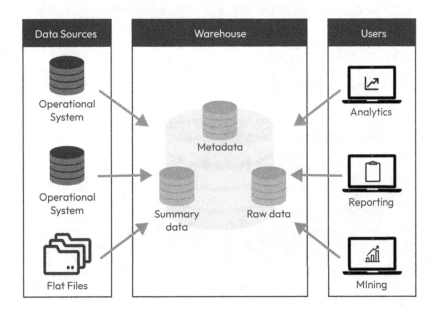

Figure 2.9 – Data sources

Timesaving

Without proper systems, this process can be time-consuming. Standardized processes and automation can save a lot of time.

Cost effective

Companies need to build streamlined processes and good practices for data wrangling, which can eventually save lots of money. At the end of the data only if the data is right, better decisions can be made ultimately resulting in good profits for the company.

Better decisions

Using data, stakeholders make important business decisions that can directly impact companies' revenue, profits, and loss. Hence, it's very important to have reliable data to make informed decisions.

Growing data

Data is produced each day and is stored in multiple different structured and unstructured formats. Data wrangling plays a key role in making sense of all these formats, connecting them, and bringing everything into one place.

Trust

With proper systems and processes in place, all the dirty data validations can be systematically handled, hence building consistency and trust in the data. Trust in data is an important aspect of decision-making and generating insights. Data wrangling helps to build that trust.

Automation

More and more companies are adopting streamlined processes and automation pipelines for data wrangling by using different data wrangling tools and moving away from the legacy approach of manually transforming the data. Real-time analytics requires data instantly, and automated data wrangling helps to generate clean and usable data within seconds.

Figure 2.10 – Data pipeline automation

Benefits of data wrangling

Let's look at some of the benefits of data wrangling:

- Data wrangling integrates huge datasets from different sources properly and makes sure data is available for analysis
- With the right automation tools and processes, data can easily be transformed into clean data after going through multiple validations to minimize noise and errors
- Data wrangling ensures that large volumes of data are processed properly with standard data-wrangling processes
- Data wrangling identifies any problems or issues with the data

Data wrangling use cases

Let's look at some example uses of data wrangling:

- Integrating data from multiple data sources into a single dataset for analysis
- Identifying missing values in the data and either imputing or deleting them
- Identifying inconsistent formats in the data and cleaning that data
- Ensuring consistent data types are maintained while combining datasets
- Adding logical checks in the data wrangling process for validation purposes
- Detecting outliers in the data and either removing them or taking some action to build consistent data

Business use cases

Here are some examples of how data wrangling can benefit a business:

- Ensuring data security is maintained
- Reducing time spent on data wrangling through automation, allowing analysts to focus on analysis
- Understanding outliers, trends, and patterns in the data
- Ensuring proper validation checks are being done as part of compliance, especially in the banking industry

Now that we have read about the importance of data wrangling, let's look at what data capture is and how data is captured in general.

Data capture

Data capture is the process of collecting information from online or electronically in structured or unstructured formats and storing it in a readable format in a table or a document as raw data. For example, weather data from a website (such as `https://www.timeanddate.com/weather/usa/boston/historic`) or the top 10 Fortune 500 companies (`https://fortune.com/fortune500/`) can be extracted from these websites and then can be stored in an Excel or CSV file.

How does data get captured?

Every day in our day-to-day life we either capture data ourselves or see data being created by us. For example, in bank transactions, we don't create that data, but data is captured and stored electronically in different database tables, which we review as our daily transaction activity or in statements. Swiping a credit card does not just pay a bill; it is also a process of creating a trail of data.

Data-capturing techniques

There are several data-capturing techniques that are used to capture data based on the type of data. For example, manual data entry is used by many businesses to capture account information, such as expenses, assets, and liabilities that will be used at the end of the year to file taxes.

Similarly, there are many different data-capturing techniques:

- Manual
- **Optical Character Recognition (OCR)**
- Barcode/QR code recognition
- Smart forms/Google Forms
- Image and video capture
- Web scraping
- Voice capture

All these data-capturing processes sound and seem easy, but each of them has its own challenges and advantages.

These are some of the challenges of data capture:

- Expensive
- Time consuming
- Misinterpretation – high rates of false positives
- Low quality

And these are some of the advantages of data capture:

- Increased data availability
- Helps avoid errors
- Increased visibility of data
- Increased confidence in data analysis results

Let's look at one of the most common ways to capture data, which is web scraping.

Web scraping

As the name suggests, web scraping is a technique in which we extract data from a website and then save it in a more useful format, such as a CSV, Excel, or JSON file, depending on the type of data we are extracting and the type of data document we are trying to create.

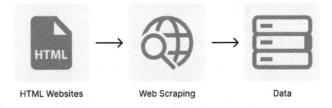

HTML Websites   Web Scraping   Data

Figure 2.11 – Web scraping process

Here are some tools that are used for web scraping:

- ParseHub
- Scrapy
- Talend
- Alteryx
- Python Beautiful Soup

Once the data gets scraped into a document in the form of an Excel or CSV file, or whatever form you require, then it can be easily manipulated and transformed into any desired form, or it can be loaded into SQL as a database table, which can be used for query building and inference deduction.

> **Note**
> Nowadays, many websites try to block web scraping because they may have confidential data that they don't want the public to have direct access to. In those cases, if the data is needed, then it's best to contact the page owner and request data access directly instead of trying web scraping.

Structured versus unstructured data

Now that we have learned about data wrangling, let's try and understand what kind of data gets extracted and stored in an Excel or CSV file.

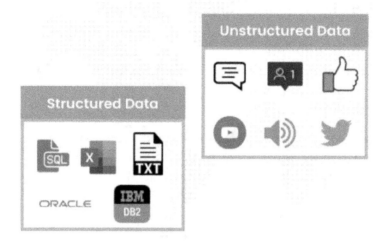

Figure 2.12 – Structured versus unstructured data

There are two different kinds of data generated from web scraping, structured and unstructured data, as described here.

Structured data

As the name suggests, this is data in an organized format that is arranged into rows and columns.

STRUCTURED DATA

Figure 2.13 – Structured data

For example, the customer data that we looked at earlier can be stored either in Excel or CSV format, or a database that follows all the rules of normalization. Each row has a unique value and can be analyzed or wrangled using **Structured Query Language (SQL)**.

Customer Id	Name	Address	Phone	Gender	Email
1	Joey	Texas	834-2345	M	JT@domain.com
2	Ron	Tennessee	987-6543	M	RT@domain.com
3	Fred	New York	876-5678	M	FN@Domain.com
4	Tom	LA	765-7654	M	TL@domain.com
5	Mary	Georgia	124-0987	F	MG@domain.com

Figure 2.14 – Customers table (structured data)

Unstructured data

Contrary to structured data, unstructured data is not organized or stored in the form of rows and columns.

UNSTRUCTURED DATA

Figure 2.15 – Unstructured data

Examples of unstructured data are customer reviews, tweets, videos, and audio. Let's say our customer **Joey** ordered Product 1 and left a rating along with some feedback comments. After one week, Joey ordered Product 2 and left a review for it along with an image showing that the product is defective. How is this feedback stored in the database? Since this is text and image data, it cannot be stored directly in a row or a column format and we need some other storage formats, such as NoSQL (a non-relational database), in a data lake or data warehouse, or in a Storage Area Network.

Let's look at the key differences between structured and unstructured data:

Structured	Unstructured
Data stored in rows and columns	Unorganized vast data
Scales slowly	Scales exponentially
Data types – int, Boolean, string	Data types – images, text, emails, messages, feedback reviews
Can be easily used by non-technical users as well	Cannot be used by non-technical users easily as it's complex and needs transformation
Example: Excel, SQL database	Example: Survey results – how many times you use Excel

Figure 2.16 – Structured versus unstructured data

Paid-for versus free data-wrangling tools

There are many different data-wrangling tools available on the market nowadays that we can use and extract relevant and meaningful data.

These are some free tools:

- **Tabula**: This tool extracts data stored in PDFs into CSV or Excel files. This tool is available for free at `https://github.com/tabulapdf/tabula`.

- **Open Refine**: This is an open source Google tool that wrangles inconsistent and messy data from one format to another and can also extend data by connecting it with the web. Using this tool is the best and easiest way to identify inconsistencies in the data and can be found in the following locations:

  - `https://openrefine.org/`

  - GitHub: `https://github.com/OpenRefine/OpenRefine`

- **R packages**: The R language has many packages, or pre-programmed functions, that help with data wrangling. Two of the most common and efficient packages in R are `dlpr` and `tidyr`:

  - Some of the functions in `dlpr` are as follows:

    - `mutate()` adds new variables that are functions of existing variables

    - `select()` picks variables based on their names

    - `filter()` picks cases based on their values

    - `summarise()` reduces multiple values to a single summary

- `arrange()` changes the ordering of the rows

- Some of the functions in `tidyr` are as follows:

You can pivot data from rows to columns and columns to rows. `pivot_longer` and `pivot_wider` are the two functions for this purpose. For example, let's use `pivot_longer` and see the output.

Here is the input:

Customer Id	Pickup	Delivery
1	10	5
2	5	10
3	15	5

Figure 2.17 – Input table

And here is the output:

Customer Id	Mode	Value
1	Pickup	10
1	Delivery	5
2	Pickup	5
2	Delivery	10
3	Pickup	15
3	Delivery	5

Figure 2.18 – Output table

- **CSVKIT**: This tool can be used when we are trying to convert a CSV file to JSON or Excel format. This can be very useful when we download data directly from SQL into CSV format and then want to convert it to JSON format.

  Let's look at an example:

```
query "select name from data where customer_age > 30" data.csv >
new.csv
```

- **SQL**: SQL can be used as a data-wrangling tool to extract and transform data stored within a database. Transactional data with several thousand rows being created daily are stored in a SQL database. This data is sometimes not clean and ready to use; it needs a lot of individual steps to prepare the data and finally make it ready for consumption. SQL syntax such as `trunc` and `trim`, and aggregate functions such as `Lead`, `Lag`, and `dense_rank()` can be used for data wrangling.

- **Mr Data Converter**: This has a user interface like an app, where you input data in one format and the result will be generated as per your selection in the dropdown, as shown in *Figure 2.19*. You can access Mr Data Converter at `http://shancarter.github.io/mr-data-converter/`.

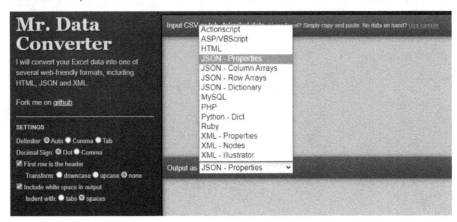

Figure 2.19 – Mr Data Converter

Let's now look at some tools that you need to pay for:

- **Talend**: This is a browser-based platform with a simple point-and-click interface. This is one of the most powerful and intelligent tools for performing data wrangling, but you need to buy a license in order to use it. You can access it at `https://www.talend.com/free-trial/`.

- **Alteryx**: I am certified to an advanced level on this tool, so I can personally 100% vouch for this tool. It is one of the best data wrangling and manipulation tools. This not only includes capabilities such as data blending and transformation, but also performing analysis such as special analysis and machine learning. This is also one of the most compatible tools for analysis with SQL.

- **Microsoft Power Query**: This is the most popular data wrangling tool and it can be used with Excel. You can directly write SQL queries or fetch data from different data sources, such as CSV files, text files, or databases, into Excel directly and then carry out data wrangling on it directly. This feature is one of the most beloved features of Excel, but it has drawbacks such as slow Excel computation and an inability to ingest large datasets into Excel.

- **Tableau**: This is a data visualization tool, but it can also prepare data using high-level data wrangling methodology. You can perform string operations such as trimming whitespaces or splitting string into two or three parts. This is a paid tool and you need a license for every user, so it becomes expensive for big organizations to use Tableau just for data wrangling.

I would like to conclude by saying that all these tools help us wrangle data faster and more efficiently and enrich it to make it ready for analysis.

Data profiling

Data profiling is the process of assessing, analyzing, and building useful summaries of data quality to decide whether the data can be used for any particular project. This is like a validation step to find any inconsistencies, inaccurate data, and missing values so that they are corrected before deciding whether to proceed with a project.

This involves creating useful summaries and visualizations that provide important metrics, such as the percentage of duplicate values in the data, the accuracy of the data, and the overall data quality score. It is a useful way to ensure the structure is correct and the relationships between different datasets are consistent. It also includes tagging different datasets, which involves assigning keywords to enhance the speed of analysis in the future so that data is easily searchable.

These are the main objectives of data profiling:

- Validating descriptive statistics of the data, such as min, max, count, and sum
- Validating different data types and lengths of the data
- Validating trends and patterns
- Performing data quality checks
- Validating different join conditions and how they perform
- Finding out metadata and assessing its quality
- Identifying different dependencies between tables

Data profiling types

In the realm of data profiling, various types include structural profiling, statistical profiling, and content profiling, each offering unique perspectives on data quality and characteristics for effective data management and analysis.

Figure 2.20 – Data profiling types

Structure discovery

This is the process of validating that the data's structure and format are consistent, that is, ensuring data is in the right format and that overall consistency is maintained. Usually in this step, data engineers validate that a given pattern matches with incoming data from either a transform job or SQL code.

For example, if a pattern of mobile numbers is known, the data engineer would match the counts of digits of the incoming phone numbers to find mismatches in the data. If the data has any inconsistencies or bad characters, it can be easily spotted in this step.

In this step, statistics such as the mean, median, mode, and standard deviation of the data are generated. These statistics can also be useful indicators for the future.

Relationship discovery

This step is important when there are multiple tables and datasets. It ensures data integrity is maintained throughout all the tables and datasets and validates relationships between tables to ensure dependencies are not lost while importing the data.

Content discovery

This step involves looking at individual records of the data to identify errors. We can validate some obvious errors in the data through this step, such as missing values, ambiguous data, and any systematic errors.

Example 1

Some values within the `Customer_Name` column are lowercase, while some values are uppercase.

ID	Customer_Name	Customer_Phone_No
1	Rahul	888-975-7468
2	Ram	988-975-7469
3	AMIR	899-975-7470
4	Atul	889-975-7471

Figure 2.21 – Customer table

Example 2

Some numbers in the `Product_Price` column have the integer data type, while some have the text data type. If we merge this data, we might not get statistically correct data.

Product_ID	Product_Name	Product_Price	Data Type
1	Pen	10	Int
2	Bat	20	Int
3	Book	30	Varchar

Figure 2.22 – Product table

Data profiling techniques

In the realm of data profiling, techniques such as data discovery, data quality assessment, and data pattern identification are employed to gain comprehensive insights into data assets, enabling organizations to enhance data understanding, reliability, and decision-making.

Column profiling

This is the process of extracting the total number of times a value occurs in each column. This can be an important way to identify the most frequent values and patterns within the data.

Cross-column profiling

This technique is used to look at the different columns to validate dependencies and keys between different columns. This step is useful to identify any valid primary keys and dependencies within the dataset, which can then be used to identify dependencies between different datasets.

Cross-table processing

This technique is used to look at different tables to identify foreign key relationships. This can be used to identify relationships between tables and can help understand how to map the data correctly.

Data profiling is important for the following reasons:

- To decide whether to go ahead with a project based on the data we have and whether it is suitable for analysis
- Validating that data quality is consistent and aligns with expectations
- Validating that the data quality foundation is strong and dependencies are not lost when moving data from one location to another

Let's have a look at some data profiling best practices:

- Get a distinct count and percentage of each column to identify errors. This can help fill those gaps and add data through inserts and updates.

- Get the percentage of zeros, null values, and blanks within the data. This can aid decisions regarding whether data should be deleted or added.

- Validating data types in each column to identify any mismatches within the data.

- Validating standard statistics of the data, such as min, max, and standard deviation.

- Validating cardinality between different datasets, such as one-to-one, one-to-many, and many-to-many.

Practical exercise

In this practical exercise, let's walk through the steps of the data-wrangling process using the provided data.

Step 1 – Discovery

CustId	Name	Addrs	Phone	Gender	Email
1	Joey	Texas	902-834-2345	M	JT@domain.com
2	Ron	Tennessee	0	/m	RT@domain.com
3	Fred	New York	902-876-5678	M	FN@Domain.com
4	Kim	LA	902-765-7654	F	TL@domain.com
5	Mary	Georgia	0	F	MG@domain.com
3	Fred	New York	902-876-5678		FN@Domain.com

Figure 2.23 – Customers table

We can see that the table holds customer information. We assume that the meanings of the different columns are as follows:

- **CustID**: This holds customer ID information

- **Name**: Customer name

- **Addrs**: Customer address

- **Phone**: Customer phone number

- **Gender**: Customer gender

- **Email**: Customer email

We first find the distinct count of each column. That will help us understand if the count matches the expected total records in each column:

```
Select count(x) from table
```

> **Note**
> x refers to a particular column/attribute in the table.

Here's an example:

```
Select count(phone) from customers
```

The output is 6.

Do this for all the columns using this query:

```
Select count(x), count(y), count(z)….. from table --
```

The preceding code will get the count for all columns.

We then extract the distinct count of the records in each column. Here is the syntax:

```
Select count(distinct x) from table
```

Here's the SQL code to extract a count of all the distinct phone numbers from the table:

```
Select distinct count(phone) from customers
```

The output is 4.

Follow this step for all the columns:

```
Select distinct count(distinct x), count(distinct y), count(distinct
z)….. from table - for all the columns to get the count
```

We can use this with the preceding query to see if there are any duplicates.

In this example, we can see that the results of query 1 and query 2 do not match. Hence there are duplicates that need to be handled.

We then use the following query to see what the distinct values in the data are:

```
Select
x,
count(distinct x)
from table
Group by x
```

This query gives us the number of times each record occurs in a column.

For example, use this query on the Customers table:

```
Select
 phone,
```

```
count(distinct phone) as count1
from table
Group by phone
```

We get the following result:

Phone	Count1
902-834-2345	1
902-876-5678	2
902-765-7654	1
0	2

Figure 2.24 – Extracting the count of distinct phone numbers

There are two observations from this result:

- We can see that **0** occurs twice in the data. This is the missing value, and it needs to be handled.

- We can also see that we have duplicate data because the phone number **902-876-5678** appears twice, and hence we could have duplicate data.

When we run the same query for the Gender column, we can see that there are bad characters in the data:

Gender	Count1
M	2
F	2
/m	1

Figure 2.25 – Extracting the Gender count

We can make two observations from the result table:

- This doesn't match the total count of records, which is 6. Hence, we can say that there is **1** null value.

- We can also clearly see there is one bad character that needs to be handled.

Step 2 – Structuring

In this step, we change the column headings to make them easy to understand as follows:

Customer_ID	Name	Address	Phone	Gender	Email

Figure 2.26 – Column headings

We also make sure that all the data types are consistent throughout the data and are in the right format. In MySQL, we can do this using the `Describe` statement:

```
Describe Customers;
```

This will give the data type of all the columns. If there are any mismatches, we need to make sure the column has the right data type (integer, string, date, and so on).

Step 3 – Cleaning

In this step, we handle errors in the data. The errors we identified in the Discovery step will be handled in this step.

Handling null values

In the earlier steps, we identified that missing values are being reflected as 0. We can convert them into nulls using the following query:

```
Update Customers
Set phone=NULL
Where phone=0
```

This is the result:

CustId	Name	Addrs	phone	Gender	Email
1	Joey	Texas	902-834-2345	M	JT@domain.com
2	Ron	Tennessee	NULL	/m	RT@domain.com
3	Fred	New York	902-876-5678	M	FN@Domain.com
4	Kim	LA	902-765-7654	F	TL@domain.com
5	Mary	Georgia	NULL	F	MG@domain.com
3	Fred	New York	902-876-5678	M	FN@Domain.com

Figure 2.27 – Customer table after handling null values

Once this step is done, we use `is null` to deal with the missing values:

```
Select count(*)
From customers
Where phone is null
```

These are the three ways to deal with missing values:

- Delete them

- Impute them (using a relevant value, such as mean, median, or mode)

- Do nothing

In this case, since each customer will have a different phone number, we can't use the first two options because we don't want to delete that customer's data. Hence, we stick to the third option and do nothing once 0 values have been converted to null.

Handling duplicate values

We can delete duplicate values using the following query:

```
With cte as (
  Select
      CustID,
      Name,
      Addrs,
      Phone,
      Gender,
      Email,
      Row_number() over (Partition by CustID,Name Order by CustID,Name
) row_num1
From Customers
)
Delete
from cte
Where row_num >1;
```

In this query, we are examining unique customer details and deleting them if the customer ID appears more than once to eliminate duplicate data.

Step 4 – Enriching

In this step, we will extract additional data from different datasets. Let's say that customer data is available from more than one data source. In this scenario, we will extract data from different data sources and make sure the data is consistent once we combine it.

Customer_ID	Name	Address	Phone	Gender	Email
1	Joey	Texas	902-834-2345	M	JT@domain.com
2	Ron	Tennessee		M	RT@domain.com
3	Fred	New York	902-876-5678	M	FN@Domain.com
4	Kim	LA	902-765-7654	F	TL@domain.com
5	Mary	Georgia		F	MG@domain.com

6	Tim	Texas	902-834-2345	M	JT@domain.com
7	Lee	California	902-834-2346	M	RT@domain.com
8	July	Washington	902-876-5678	F	FN@domain.com
9	Julia	LA	902-765-7654	F	TL@domain.com
10	Sam	Georgia	902-723-7827	F	MG@domain.com

Figure 2.28 – Customer table after enrichment

> **Note**
> Before you extract the data from other data sources, one thing to remember is that it has to go through the preceding three steps before the data is merged with the current data.

Step 5 – Validating

In this step, we will validate different business logic by building a few custom calculations and also re-validate the previous queries to check that the data is consistent and as expected without any errors. If we still observe any errors, we will eliminate them in this step.

Step 6 – Publishing

Once we have confirmed that the preceding steps have been followed, we finally publish the data and make it available for analysis. After this step, the data is in a clean, consistent, and usable format, and it can be used for analysis. The clean data can be used for creating dashboards and predictive models after this step.

Summary

In conclusion, this chapter has provided an overview of the essential concepts related to data wrangling. We explored the definition of data wrangling and its fundamental steps, which are Discovery, Structuring, Cleaning, Enriching, Validating, and Publishing. We discussed the importance and benefits of data wrangling, along with its diverse use cases. We also touched upon data-capturing and web-scraping techniques, as well as the commonly used data-wrangling tools in the industry. Additionally, the significance of data profiling was discussed, and we demonstrated the application of data-wrangling techniques with a practical example. This comprehensive exploration of data wrangling equipped you with valuable insights and knowledge to effectively manage and transform data for optimal decision-making and analysis. Now that we have a clear understanding of data wrangling and its benefits, in the next chapter, we will start wrangling data with string data types.

Part 2:
Data Wrangling Techniques
Using SQL

This part includes the following chapters:

3

Data Wrangling on String Data Types

Most business data is stored in the form of strings, which are sequences of characters that store information, and SQL has multiple built-in functions that can slice and dice this data to meet a specific analysis requirement. For example, we may want to extract the month information from a date column. The goal of this chapter is to help you gain hands-on experience of extracting meaningful information after data wrangling on string data.

In this chapter, we will cover the following main topics:

- SQL data types
- The string data type and its use case
- Exploring SQL string functions

SQL data types

Each value stored in a database has a fixed data type that defines its properties, storage size, and what data size that variable can hold. Every time we perform data manipulation in SQL, we need to ensure that the data type is compatible with the manipulation operation.

Data types are categorized into the following six types, which are explained in more detail here:

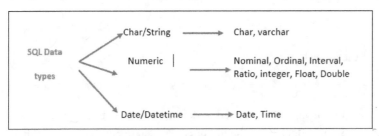

Figure 3.1 – SQL data types

Numeric data types

Overall, there are four different numerical data types that differ by the level or type of information we are trying to store, such as the number of characters/decimal points.

Nominal data type

Numbers such as phone numbers and customer IDs use the nominal data type because number operations such as `add`, `subtract`, and `compare` cannot be performed on this data type. They are for informational purposes rather than analysis. Nominal data is also called qualitative or categorical.

> **Note**
>
> Only calculations based on the frequencies or percentages of occurrence are valid on the nominal data type.

For example, we could record marital status using the following code:

```
single = 1, married = 2, divorced = 3, widowed = 4
```

Here is another coding system that is just as valid as the previous one:

```
Single = 7, married = 4, divorced = 13, widowed = 1
```

Ordinal data type

As the name suggests, this data defines the order of data – for example, survey results ranking user satisfaction such as happy, sad, average, and so on. Again, we cannot perform numerical operations on ordinal data but this data can be used to perform analysis such as sentiment analysis. Ordinal data resembles nominal data, but the difference is that the order of their values has meaning.

> **Note**
>
> The only permissible calculations are those involving a ranking process for ordinal data types.

For example, we can record student evaluations as follows:

```
Poor = 1, Fair = 2, Good = 3, Very good = 4, Excellent = 5
```

We can also assign the following codes:

```
Poor = 6, Fair = 18, Good = 23, Very good = 45, Excellent = 88
```

As you can see, any kind of value can be used to define the categories, but we need to ensure that for our choice of codes, the order must be maintained. We can use any set of codes that are in order.

Interval data type

This numerical data type stores numerical data that is an interval between two different values. An example could be an interval in years, hours, minutes, and seconds.

For example, between 10:00 and 12:30 is an interval of 02:30 (2 hours and 30 minutes). Interval values can be YEAR, MONTH, DAY, HOUR, MINUTE, SECOND.

Following is an example of a table using interval data types:

```
CREATE TABLE test_interval (
id     DECIMAL PRIMARY KEY,
col1   INTERVAL YEAR TO MONTH,
col2   INTERVAL DAY TO SECOND(6) -- an interval with 6 digits after
the  decimal of seconds);
```

Ratio value

Ratio values, as the name suggests, represent a proportion or fraction of a whole.

A few examples of data that can be represented using ratio values are income, crime rate, and age ranges in a survey response (e.g., 18-24, 24-30, or 30+).

These numbers in SQL can be further sub-categorized as *exact or approximate* in terms of data types.

> **Note**
> Exact data types are integer, numeric, and decimal, while approximate data types are float and double.

Let's look at the exact data types:

- **INTEGER**: Used to store whole numbers without decimal places.
- **NUMERIC**: Used to store numbers with a fixed precision and scale. Precision represents the total number of digits, and scale represents the number of decimal places, for example, NUMERIC(precision, scale).
- **DECIMAL**: Similar to NUMERIC, this is used to store numbers with a fixed precision and scale, for example, DECIMAL(precision, scale).

And now let's see the approximate data types:

- **FLOAT**: Used to store floating-point numbers with variable precision. It can store a wide range of values, but the precision can vary. For example, FLOAT(precision). The FLOAT data type typically requires 4 bytes or 8 bytes of storage space, and typically, it provides a precision of about 6 to 9 decimal digits.

- **DOUBLE**: This is also used to store floating-point numbers with variable precision, but typically allows a larger range of values than FLOAT. It typically offers a precision of about 15 to 17 decimal digits.

> **A few points to note**
>
> Both numeric and decimal data types hold two parts of a numerical value called precision and scale.
>
> Precision defines the number of digits in a number, and scale defines the number of digits to the right of a decimal point.
>
> For example, the number 1234.56 has a precision of 6 and a scale of 2 since there are only 2 decimal digits. The default maximum precision in SQL is 38.

In SQL, the DDL for the precision and scale will look like this:

```
CREATE TABLE customer (
Customer_id int,
credit decimal(8,2) );
```

Here, the credit column will be able to store a number with up to 10 digits.

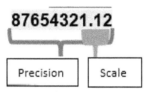

Figure 3.2 – Credit column

> **The key thing to note here**
>
> The precision limit of the numeric data type should always be kept in mind before performing any manipulation using SQL because otherwise, it will create errors while running the query and cause inconsistent data fetch issues.

Integer data type

Integer data types store numbers that are whole numbers and do not have any decimal values. Examples of integer data types are smallint, bigint, and int. The most commonly used integer data type is int, which stores up to 4 bytes of data, whereas smallint stores 2 bytes and bigint stores up to 8 bytes of data.

> **Tip**
> Knowing the storage size is important because it helps keep the database space utilization in check, keeps the database tables small, and helps with the speed of read operations.

The DDL code for integer data is as follows:

```
CREATE TABLE customer (
Customer_id int,
credit integer
);
INSERT INTO customer (Customer_ID, credit) VALUES(10,12345.5);
```

We know that the credit column is of the integer data type so it cannot store a decimal value, but when decimal values are ingested into an integer column, SQL processes them as whole numbers by rounding off the decimal digits. For the customer with ID 10, the credit value saved will be 12346 as per the rounding logic. This is one scenario in which the data that is being loaded into the table should be carefully audited because many such rows of data will skew the data being stored in the tables and will also affect any analysis being carried out on such data because it's no longer an accurate dataset.

Float data type

Two of the most commonly used key float data types that store the greatest amount of digits for decimals in ANSII SQL are FLOAT and DOUBLE. DOUBLE holds up to 64 bits of information, so usually, it's the best data type for storing decimal data, especially when manipulations are performed, which can cause data size to grow many times larger than the original stored value, such as property value. FLOAT or double data types should be used when the data type is being built for scientific or engineering applications.

The DDL for float data is as follows:

```
CREATE TABLE customer (
Customer_id int,
credit float
);
INSERT INTO customer (Customer_ID, credit) VALUES(10,12345.57654);
```

Here Customer_id '10' holds a Credit value of '12345.57654':

INT	FLOAT
Customer_id	Credit
10	12345.57654

Figure 3.3 – Customer table

Date and time data types

Datetime data types are `DATE` and `TIME`, which hold different formats of date values.

Date

With this data type, you can store simple calendar dates in the formats YYYY-MM-DD or MM-DD-YYYY from Jan 1, 1753, to Dec 31, 9999.

The DDL for date data is as follows:

```
CREATE TABLE customer (
Customer_id int,
Transaction_date date, -- date type column declaration
Credit float
);
INSERT INTO customer (Customer_ID, transaction_date, credit)
VALUES(10,11-10-2022,12345.57654);
INSERT INTO customer (Customer_ID, transaction_date, credit)
VALUES(11,2022-10-12,12345.57654);
Customer_id 10 received credit for amount of 12345.57654 on November
10th, 2022
```

INT	FLOAT	DATE
Customer_id	Credit	transaction_date
10	12345.57654	11/10/2022
11	12345.57654	2022/10/12

Figure 3.4 – Storing calendar dates

Time

The time data type is used to store the time of day. Time columns store up to six digits of precision. An example value is 12:30PM.

The DDL for time data is as follows:

```
CREATE TABLE customer (
Customer_id int,
Transaction_date date,
Transaction_time time,
Credit float
);
INSERT INTO customer (Customer_ID, transaction_date, transaction_time,
credit) VALUES(10,11-10-2022,12345.57654);
```

A credit of `12345.6` for `Customer_id=10` was made on 10th November at 12:30 P.M.

INT	DATE	TIME	FLOAT
Customer_id	transaction_date	transaction_time	Credit
10	11/10/2022	12:30:07	12345.6

Figure 3.5 – Storing time

String data type

String data types are used to represent sequences of characters (such as letters, numbers, and symbols) in programming languages.

Char

`Char` stands for character. In SQL, the maximum length for this data type is 8,000 characters.

The DDL for char data is as follows:

```
CREATE TABLE customer (
Customer_id int,
Customer_name char(60) -- Here 60 defines the maximum length of the
value which can be stored in in this column
);
INSERT INTO customer (customer_ID, customer_name) VALUES(10,'JK');
Here for customer _id='10' the name is 'JK'
```

INT	Char
Customer_id	Customer_name
10	JK

Figure 3.6 – Storing characters

Varchar

Varchar stands for *variable-length strings*. The maximum length for varchar is 8,000 characters.

The DDL for this data type is as follows:

```
create table test(
col1 varchar(120) -- This means it can hold 120 single-byte
characters, 60 double-byte characters, 40 three-byte characters, or 30
4 byte characters.
);
```

Practical exercise – data type

Information about a book's readers is of interest to both the publisher and the book's advertisers. A survey of readers asked respondents to complete the following:

a. Age

b. Gender

c. Marital status

d. Number of magazine subscriptions

e. Annual income

f. Rate the quality of our magazine: excellent, good, fair, or poor

Identify the resulting data type for each item.

SQL string functions

SQL string functions are functions that perform operations on string data types. These functions are often used in SELECT, INSERT, UPDATE, and DELETE statements. These string functions allow you to manipulate character strings in SQL.

Some common tasks that you can use string functions for are as follows:

- Extracting a substring from a string
- Concatenating two or more strings
- Finding the length of a string
- Converting a string to uppercase or lowercase
- Trimming leading or trailing whitespace from a string

RIGHT()

This function returns a specific number of characters from the rightmost end of the string.

This has two parts to the argument:

- The string
- The number of characters that should be extracted from the rightmost end of the string

The syntax is as follows: RIGHT(string, number_of_chars).

Let's look at an example:

```
SELECT RIGHT('This is String Right function',14)
```

Result:

This will output `Right function` – in essence, the 14 characters from the rightmost end of the string.

Let's look at an example using the customer table:

```
Select right(Name,4) as output from customer;
```

Customer Id	Name	Address
1	Joey Ramsay	Texas
2	Ron Welch	Tennessee
3	Fred Smith	New York
4	Tom Harris	LA
5	Mary Twain	Georgia

Output
Msay
Elch
Mith
Rris
Wain

Figure 3.7 – Output from using the RIGHT() function

LEFT()

This function returns a specific number of characters from the left end, or start, of the string.

This has two parts to the argument:

- A string stating the column from which the characters are to be extracted
- The number of characters that should be fetched from the left side of the string

Here is the syntax: LEFT(`string, number_of_chars`).

And here is an example:

```
SELECT left('This is String Left function',4)
```

Result:

This will output `This` – in essence, the first 4 characters from the left side of the string.

Let's look at an example using the customer table:

```
Select left(Name,4) as output from customer;
```

Customer Id	Name	Address
1	Ramsay	Texas
2	Ron Welch	Tennessee
3	Fred Smith	New York
4	Tom Harris	LA
5	Mary Twain	Georgia

Output
Ram
Ron
Fred
Tom
Mary

Figure 3.8 – Output from using the LEFT() function

LEN()

This function returns the length of the string, that is, the total number of characters in the string, with spaces included. This function has one argument: the string or the variable holding the string value.

Here is the syntax:

```
SELECT LEN(string)
```

And here's an example:

```
SELECT len('This is String length function')
```

Result:

This will output 30 since the string contains 30 characters.

Let's look at an example using the customer table:

```
Select len(Name) as output from customer;
```

Customer Id	Name	Address
1	Joey Ramsay	Texas
2	Ron Welch	Tennessee
3	Fred Smith	New York
4	Tom Harris	LA
5	Mary Twain	Georgia

Output
11
9
10
10
10

Figure 3.9 – Output from using the LEN() function

TRIM()

This function is used to remove space characters and any other characters from the start or end of the string. It also removes leading and trailing spaces from the string by default.

Here is the syntax:

```
TRIM([characters FROM ]string)
```

And here is an example:

```
SELECT TRIM('  This is String TRIM function    ');
```

Result:

`This is String TRIM function` – Here, the leading and trailing spaces are removed by the `TRIM` function.

RTRIM()

This function is used to trim all the *whitespace* from the *right* side of the string. This has one argument, which is a string or a variable holding the string value.

These are common whitespace characters:

- **Space**: The standard space character that represents a single empty space
- **Tab**: A character used to create a horizontal indentation in text or data
- **Newline (line feed)**: A character used to indicate the end of a line and the start of a new line

Here is the syntax:

```
SELECT RTRIM(string)
```

And here's an example:

```
select len('      This is String RTRIM function      ')
```

Result:

`'This is String RTRIM function'` The trailing whitespace on the right side of the string is deleted by the `RTRIM` function.

LTRIM()

This function is used to trim all the whitespace from the left side of the string. This has one argument, a string or a variable holding the string value.

Here is the syntax:

```
SELECT LTRIM(string)
```

And here is an example:

```
select len('       This is String RTRIM function        ')
```

Result:

'This is String LTRIM function' – The initial whitespace is removed by the LTRIM function.

RPAD()

The RPAD function takes a text value and pads it on the right by adding a specified number of extra characters to the right of the value.

Here is the syntax:

```
RPAD(string, length, rpad_string)
```

The parameters of the SQL RPAD function are as follows:

- string (mandatory): This is the text value or text expression that you want to pad.
- length (mandatory): This is the total length value that the expression will be padded to.
- rpad_string (optional): This is the character or set of characters to use to pad the string value. The default value is a single space.

Here is an example:

```
Select RPAD('6',4,'0') as new_column;
```

Result:

60000 – Here, four 0s are added to the right of the character '6'.

And here's another example:

```
SELECT
first_name,
RPAD(first_name, 10, '*') AS padded_value
FROM customers;
```

FIRST_NAME	PADDED_VALUE
John	John******
Sally	Sally*****
Steve	Steve*****
Adam	Adam******

Figure 3.10 – Output from using the RPAD() function

LPAD()

This function adds left padding with a given symbol to make a string of a certain fixed size. This function's syntax is the same as RPAD with similar parameters.

Here's an example:

```
Select LPAD('8',4,'0') as new_column;
```

This is the result:

```
0008
```

Let's see another example with LPAD and RPAD in *one statement*, with different characters to show you how they are added:

```
SELECT
last_name,
RPAD(LPAD(last_name, 10, '#'), 15, '*') AS padded_value
FROM customers;
```

Here's the result:

LAST_NAME	PADDED_VALUE
Smith	#####Smith*****
Jones	#####Jones*****
Brown	#####Brown*****
Cooper	####Cooper*****

Figure 3.11 – Output from using the LPAD() function

These are the characteristics of the LPAD and RPAD functions:

- If the length value is shorter than the length of expr, then expr is trimmed to the defined length

- If `pad_expr` is not specified, a single space is used
- The return data type is `TEXT`

> **Note**
>
> It is possible that the `LPAD/RPAD` syntax might not work. In that case, check for the following mistakes:
>
> - The length might be shorter than the expression
> - You might need to use `RPAD` instead of `LPAD`
> - You might be trying to insert the wrong data type or size of data into a column
> - LPAD/RPAD will not work with NULL values

Some invisible characters, such as null and new line, can be introduced through various means, such as copying and pasting text from different sources, encoding issues, or data import/export processes. When invisible characters are present, they can affect the behavior of string manipulation functions such as LPAD/RPAD. For example, if an invisible character is at the beginning or end of a string, it may not be recognized or handled as expected by these functions, leading to unexpected results.

REPLACE()

This function is used to replace all occurrences of specific characters in a string.

This function has three arguments:

- The string in which we want to replace the character
- The character that we want to replace
- The character that we want to replace it with

Here is the syntax:

```
REPLACE(string, pattern, replacement)
```

The parameters of the SQL `REPLACE` function are as follows:

- **String**: The input string value on which the `REPLACE` function has to operate
- **Pattern**: The substring to evaluate and a reference position to the replacement field
- **REPLACEMENT**: Replaces the specified string or character value of the given expression

> **Note**
>
> The SQL REPLACE function performs comparisons based on the collation of the input expression.

Here's an example:

```
select replace('String Function ','n','$') -- This will replace all
the occurrences of 'n' by a '$'.
```

This is the result:

```
Stri$g Fu$ctio$
```

Let's look at another example:

```
SELECT Name,
       value,
       REPLACE (REPLACE (REPLACE(value, 'A', '5'), 'C', 9), 'D', 4) as
new_value
FROM datatable;
```

This is the result:

Name	value	new_Value
Smith	AB	5B
Jones	ABC	5B9
Brown	ABCD	5B94

Figure 3.12 – Output from using the REPLACE() function

Interesting read

Let's try to convert a varchar field to a number using the following table. Note that there is a set of common characters inside each varchar field that needs to be removed in order for it to be successfully converted to numeric.

Product_Id	Selling_Price(Varchar)
1	$4.50
2	£ 5.60
3	£6.70
4	7.00
5	#N/A
6	$8

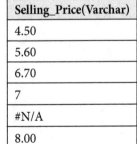

Selling_Price(Varchar)
4.50
5.60
6.70
7
#N/A
8.00

Figure 3.13 – Output from using the REPLACE() function

We need to remove the following strings from the `Selling_Price` field:

```
'.00'
'£'
'n/a'
'$'
'#N/A'
```

This can be achieved using nested `replace` functions, as follows:

```
select replace(replace(replace(col, '$', ''), '£', ''), 'n/a', '')
```

Here, the first $ will be replaced with a blank, then £ and n/a will be removed.

REVERSE()

This function is used to reverse the string. It needs one argument, which is the string to be flipped right to left.

Here are some characteristics of this function:

- This function is used to reverse the string provided
- This function accepts strings as parameters
- This function always returns strings
- This function can also take a set of integers and reverse it
- This function can even reverse float values

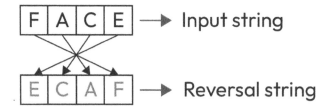

Figure 3.14 – REVERSE function

Here is the syntax:

```
REVERSE(string)
```

And here is an example:

```
Select reverse ('String')
```

This is the result:

```
gnirtS
```

SUBSTRING()

This function returns a part of the string. It needs three arguments. One is a string, the second is the starting index, and the third is the ending index.

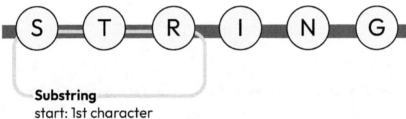

Figure 3.15 – Substring function

This is the syntax:

```
SUBSTRING(string, start, length)
```

Here's an example:

```
Select substring('This is String function',3,7)
```

This is the result:

```
is is
```

Let's look at another example:

Input	Output	Explanation
`select substring('hello world',1,5)`	hello	Extracted the 5 characters from the leftmost end of the string
select substring('hello world', -2,5)	he	Here the request is to start from the -2 location, which does not exist in the string, so only the first two characters are extracted as the length specified is 5.

Figure 3.16 – Output from the SUBSTRING() function

CAST()

This function is used to convert a given value into another data type.

This is the syntax:

```
CAST (EXPRESSION AS Data_ Type[(Length)]
```

These are the arguments:

- **Expression**: This is a valid expression in which we want to convert a data type in SQL.
- **Data_type**: This is the data type to which we want to convert the expression.
- **Length**: This is an optional parameter of the integer type. We can use the setting to define the length of any targeted data type.

Here is an example:

```
Select CAST('2022-12-12' as VARCHAR) as date_varchar    -- Saves the
date value as a varchar data type
```

Let's consider the following example to understand the table that holds satisfaction scores::

Column Name	Date Type
product Id	Integer
First_Name	char(20)
Score	Float

Figure 3.17 - Example of a satisfaction table with scores

The table contains the following rows:

product Id	First_Name	Score
1	Jami	93.2
2	Bob	89.9
3	Mani	70
4	Shawn	130.2

Figure 3.18 – Rows in the table

Here, we are using the CAST function to convert the Score column from FLOAT to INTEGER:

```
SELECT First_Name,CAST (Score AS Integer) Int_Score FROM Satisfaction;
```

Here's the result:

First_Name	Score
Jami	93
Bob	90
Mani	70
Shawn	130

Figure 3.19 – Output after using the CAST() function

As you can see in the preceding table, the numbers after the decimal point are truncated due to casting.

CONCATENATE()

This function is used to concatenate two or more strings into a single value. This function requires at least two arguments or else it gives an error.

Here is the syntax:

```
CONCAT(string1, string2, ...., string_n)
```

And here is an example:

```
Select concatenate('This','is','an','example') as variable -- This
will create a new value and store in the column variable
```

This is the result:

```
This is an example
```

CONCATENATE_WS()

This function is used to add two or more strings with a separator.

Here is the syntax:

```
CONCAT_WS(separator, string1, string2, ...., string_n)
```

And here is an example:

```
SELECT CONCAT_WS('-','This','is','an','example') -- This will add '-'
separator after every string
```

This is the result:

```
This-is-an-example
```

> **CONCAT versus CONCAT_WS**
>
> The basic difference between them is that the CONCAT_WS() function can do concatenation along with a separator between strings, whereas the CONCAT() function has no concept of the separator.

UPPER function

The UPPER function converts all string values to uppercase. The UPPER function is very useful because data can come in different formats before being loaded into a table. Let's say we are analyzing real-time feedback data that is filled in by the customer, and the customer may input data in different formats, which means that the data could be uppercase or lowercase or both. UPPER ensures that dirty data is filtered out and data is converted into a consistent format.

Here is the syntax:

```
SELECT UPPER(text)
```

There are some practical use cases of UPPER:

- To make the data consistent across the column.
- To query case-sensitive data in a column. For example, let's say we have a column with two values, MIKE and mike, and we only want to pull those that are uppercase.

Let's look at a few examples.

Example 1

Let's use the UPPER function with lowercase string values:

```
Select Upper('this is data wrangling for string') As column1_new from
test_table;
```

Here is the result:

column1	column1_new
this is data wrangling for string	THIS IS DATA WRANGLING FOR STRING

Figure 3.20 – Output after using the Upper() function with lowercase string values

Example 2

Let's use the Upper function with mixed case string values:

```
Select Upper('This IS data wrangling for string') As column1;
```

Result:

column1	column1_new
This IS data wrangling for string	THIS IS DATA WRANGLING FOR STRING

Figure 3.21 – Output after using the Upper() function with mixed case string values

Example 3

Let's use the Upper function with a select statement in the following customers table.

We can use upper to convert the string values to uppercase, as follows:

```
Select upper(Address) from customers
```

Customer_ID	Name	Address	Phone	Gender	Email
1	Joey	Texas	902-834-2345	M	JT@domain.com
2	Ron	TENNESSEE		M	RT@domain.com
3	Fred	New York	902-876-5678	M	FN@Domain.com
4	Kim	LA	902-765-7654	F	TL@domain.com

Figure 3.22 – Using the Upper() function to get uppercase string values

This is the result:

Customer_ID	Name	Address	Phone	Gender	Email
1	Joey	TEXAS	902-834-2345	M	JT@domain.com
2	Ron	TENNESSEE		M	RT@domain.com
3	Fred	NEW YORK	902-876-5678	M	FN@Domain.com
4	Kim	LA	902-765-7654	F	TL@domain.com

Figure 3.23 – Output after using the Upper() function to get uppercase string values

Example 4

Let's use `upper` with an `Update` statement:

```
Update Customers Set Name=upper(Name)
```

As we can see, we can permanently update the values in the `Name` column to uppercase:

Customer_ID	Name	Address	Phone	Gender	Email
1	JOEY	Texas	902-834-2345	M	JT@domain.com
2	RON	TENNESSEE		M	RT@domain.com
3	FRED	New York	902-876-5678	M	FN@Domain.com
4	KIM	LA	902-765-7654	F	TL@domain.com

Figure 3.24 – Converting the Name column values to uppercase

We can also use `upper` to update for a specific condition using a `Where` clause:

```
Update Customers Set Address= UPPER(Address) Where Name='FRED'
```

Customer_ID	Name	Address	Phone	Gender	Email
1	JOEY	Texas	902-834-2345	M	JT@domain.com
2	RON	TENNESSEE		M	RT@domain.com
3	FRED	NEW YORK	902-876-5678	M	FN@Domain.com
4	KIM	LA	902-765-7654	F	TL@domain.com

Figure 3.25 – Using the WHERE clause

Example 5

Another common practical use of `upper` is to use it to filter

Customer_ID	Name	Address	Phone	Gender	Email
1	Joey	Texas	902-834-2345	M	JT@domain.com
2	Ron	TENNESSEE		M	RT@domain.com
3	Fred	New York	902-876-5678	M	FN@Domain.com
4	Kim	LA	902-765-7654	F	TL@domain.com

Figure 3.26 – Customer table

Let's use the `upper` function in filter conditions as follows:

```
Select * From Customers Where upper(Address)='Tennessee'
```

This is the result:

Customer_ID	Name	Address	Phone	Gender	Email
2	Ron	TENNESSEE		M	RT@domain.com

Figure 3.27 – Output from using the Upper function in filter conditions

LOWER function

The `LOWER` function converts a string into lowercase. It works the same way as the `upper` function. But it should again be used with caution to prevent the unnecessary alteration of data. Let's say you have a column called `ProductName` that stores the names of products. If the product names were entered with a specific casing, using the `LOWER` function would alter the original case. In this case, it's important to preserve the original data as entered.

Let's look at a few examples of using the `LOWER` function.

Here is the syntax:

```
SELECT LOWER(text)
```

Example 1

Use the `LOWER` function with mixed-case string values:

```
Select LOWER('This IS data wrangling for string') As column1_new from test_table;
```

Here is the result:

column1 (before)	column1_new
This IS data wrangling for string	this is data wrangling for string

Figure 3.28 – Output from using the LOWER function with mixed case string values

Example 2

Use LOWER with a SELECT statement.

Customer_ID	Name	Address	Phone	Gender	Email
1	Joey	Texas	902-834-2345	M	JT@domain.com
2	Ron	TENNESSEE		M	RT@domain.com
3	Fred	New York	902-876-5678	M	FN@Domain.com
4	Kim	LA	902-765-7654	F	TL@domain.com

Figure 3.29 – Customer table

Let's try this example as follows:

```
Select lower(Address) from customers
```

This is the result:

Customer_ID	Name	Address	Phone	Gender	Email
1	Joey	texas	902-834-2345	M	JT@domain.com
2	Ron	tennessee		M	RT@domain.com
3	Fred	new york	902-876-5678	M	FN@Domain.com
4	Kim	la	902-765-7654	F	TL@domain.com

Figure 3.30 – Output from using LOWER with a SELECT statement

Example 3

Use LOWER with an Update statement:

```
Update Customers Set Name=lower(Name)
```

As you can see, we can permanently update the values in the Name column to lowercase:

Customer_ID	Name	Address	Phone	Gender	Email
1	joey	texas	902-834-2345	M	JT@domain.com
2	ron	tennessee		M	RT@domain.com
3	fred	new york	902-876-5678	M	FN@Domain.com
4	kim	la	902-765-7654	F	TL@domain.com

Figure 3.31 – Updating the Name column to lowercase

Example 4

Use the `lower` function in a `Where` clause:

```
Select * From Customers Where lower(Address)='new york'
```

This is the result:

Customer_ID	Name	Address	Phone	Gender	Email
3	FRED	new york	902-876-5678	M	FN@Domain.com

Figure 3.32 – Output from updating the Name column in to lowercase

INITCAP function

The `INITCAP` function is used to capitalize the first letter of each word.

Here is an example:

```
Select INITCAP('this is data wrangling for string')
```

This is the result:

```
This Is Data Wrangling For String
```

> **Note**
> The `INITCAP` function works only in Oracle and Postgres.

INSTR function

The `INSTR` function returns the position of the first occurrence of a substring within a string.

This is the syntax:

```
INSTR(string1, string2)
```

It takes the following parameters:

- `string1`: The string that will be searched.
- `string2`: The string to search for in `string1`. If `string2` is not found, 0 is returned.

Summary

This brings us to the end of the chapter. We have learned that SQL string functions are useful for manipulating and working with string data in SQL statements. These functions can be used to concatenate strings, find the length of a string, trim leading and trailing spaces, extract a portion of a string, and convert the case of a string. We also learned how important it is to be familiar with the string functions available in SQL and to understand the syntax and use of each function to effectively work with string data in your database.

In the next chapter, we will learn about null values, which are a very common occurrence in a SQL database, and it is important to handle them appropriately when doing data wrangling. A null value represents missing or unknown data, and it can cause issues if not properly dealt with.

Practical exercises

Let's now practice what we've learned in this chapter with some exercises.

Practical exercise 1

Let's look at multiple SQL string functions in one SQL query:

```
SELECT LENGTH(last_name) as "Length",
       CONCAT(first_name, ' ', last_name) as "Full Name",
       SUBSTRING(first_name, 1, 1) as "Initial",
       REPLACE(email, '@example.com', '') as "Username",
       UPPER(last_name) as "Last Name (Uppercase)"
FROM users
WHERE LOWER(first_name) = 'john';
```

This statement selects the length of the `last_name` column, the concatenation of the `first_name` and `last_name` columns as `"Full Name"`, the first character of the `first_name` column as `"Initial"`, the `email` column with the `@example.com` domain removed as `"Username"`, and the `last_name` column in uppercase as `"Last Name (Uppercase)"` for all rows where the `first_name` column is `'john'` (case-insensitive).

Practical exercise 2

Here, we use string functions in SELECT and WHERE clauses:

```
SELECT LENGTH(name), UPPER(name), SUBSTRING(name, 1, 3)
FROM users
WHERE TRIM(email) = 'packt.wrangling@example.com';
```

This example selects the length of the name column, the uppercase version of the name column, and the first three characters of the name column for all rows where the email column, after leading and trailing whitespace is removed, is packt.wrangling@example.com.

Practical exercise 3

In this exercise, we will use SQL string functions in a SELECT statement.

Suppose you have a table called employees with the following structure:

employee_id	first_name	last_name
1	Sui	San
2	Jane	Doe
3	Bob	Johnson

Figure 3.31 – employees table

Here is a SELECT statement that uses some string functions:

```
SELECT employee_id,
   CONCAT(first_name, ' ', last_name) as full_name,
   LENGTH(first_name) as first_name_length,
   UPPER(last_name) as last_name_uppercase
FROM employees;
```

This returns the following results:

employee_id	full_name	first_name_length	last_name_uppercase
1	Sui San	3	SAN
2	Jane Doe	4	DOE
3	Bob Johnson	3	JOHNSON

Figure 3.31 – Output

Here, the SELECT statement uses the CONCAT function to concatenate the first_name and last_name columns into a full_name column. It also uses the LENGTH function to get the length of the `first_name` column and the UPPER function to convert the `last_name` column to uppercase.

Practical exercise 4

Let's use SQL INSTR functions in a SELECT statement.

Suppose you have a table called `customers` with the following data:

first_name	last_name	Email
Sui	San	`sui.san@gmail.com`
Jane	Doe	`jane.doe@gmail.com`
null	Johnson	Null

Figure 3.32 – customers table

Here is a SQL query that wrangles the data using string functions:

```
SELECT

  UPPER(first_name) as first_name,

  LOWER(last_name) as last_name,

  CONCAT(first_name, ' ', last_name) as full_name,

  LENGTH(email) as email_length,

  SUBSTRING(email, 1, INSTR(email, '@')-1) as username

FROM customers;
```

This returns the following results:

first_name	last_name	full_name	email_length	username
Sui	San	Sui San	18	sui.san
JANE	doe	Jane Doe	17	jane.doe
null	johnson	null johnson	Null	Null

Figure 3.33 – customers table

4

Data Wrangling on the DATE Data Type

Most transactional data has some or other form of date-type data that holds meaningful information, especially when doing analysis on historical data or data modeling. In SQL, the DATE data type is used to store date values. Dates are stored in the format YYYY-MM-DD, where YYYY represents the year, MM represents the month, and DD represents the day. For example, January 5, 2020 would be stored as 2020-01-05. The DATE data type is supported by most SQL implementations, including MySQL, PostgreSQL, and SQLite. In addition to the DATE data type, most SQL implementations also support additional data types for storing time or timestamp information, such as TIME and TIMESTAMP.

For example, extracting just the month information from a date column in a table. The goal of this chapter is to help you achieve hands-on experience in how to extract meaningful information after data wrangling on the DATE data type.

In this chapter, we will cover the following topics:

- The SQL DATE data type functions
- The DATE data type and its use case
- Exploring SQL DATE functions

SQL DATE data type functions

The DATE data type typically takes 4 bytes of storage in a database and can hold date values with a range of 1000-01-01 to 9999-12-31. The TIMESTAMP data type is similar to the DATE data type, but it also includes information about the time to a precision of one second. The format used to store timestamp values is typically YYYY-MM-DD HH:MM:SS.

There are several date functions provided by the SQL engine to manipulate the date and time information stored in date or DateTime columns. Let's look at each of them in detail.

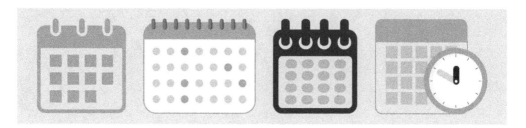

Figure 4.1 – DateTime functions

EXTRACT

The SQL EXTRACT function is used to extract a part of a date or a timestamp value. The function takes two arguments: the first is the part of the date or timestamp that you want to extract (such as the year, month, or day), and the second is the date or timestamp value that you want to extract from. The syntax for the EXTRACT function is as follows:

```
EXTRACT(date_part FROM date_or_timestamp)
```

Here, date_part is the part of the date or timestamp that you want to extract.

For example, you can use YEAR to extract the year, MONTH to extract the month, or DAY to extract the day. The date_or_timestamp is the date or timestamp value that you want to extract from.

Let's look at some examples.

One business use case for using the EXTRACT() function in SQL would be to analyze sales data. For example, a company may want to extract information about sales by year, month, day, or hour in order to gain insights into their sales trends and patterns:

- Here is the code for extracting sales by year:

```
SELECT EXTRACT(YEAR FROM order_date) AS year, SUM(total_sales)
FROM orders
GROUP BY year
ORDER BY year;
```

This query will give the company a summary of total sales for each year.

- Here is the code for extracting sales by month:

```
SELECT EXTRACT(MONTH FROM order_date) AS month, SUM(total_sales)
FROM orders
GROUP BY month
ORDER BY month;
```

This query will give the company a summary of total sales for each month.

- Here is the code for extracting sales by day of the week:

```
SELECT EXTRACT(DOW FROM order_date) AS day_of_week, SUM(total_
sales)
FROM orders
GROUP BY day_of_week
ORDER BY day_of_week;
```

This query will give the company a summary of total sales for each day of the week.

- Here is the code for extracting sales by the hour:

```
SELECT EXTRACT(HOUR FROM order_date) AS hour, SUM(total_sales)
FROM orders
GROUP BY hour
ORDER BY hour;
```

This query will give the company a summary of total sales for each hour of the day.

By using the EXTRACT() function to extract different parts of the date or time, the company can analyze its sales data in a variety of ways and gain insights into patterns and trends that could inform its business decisions.

Case scenario

The EXTRACT() function is very useful for extracting a part of a date value, such as the day, month, year, or time, which is very common when working in a day-to-day analytics role.

Consider a table named `sales` with the following structure:

sale_id	sale_date	Product	amount
1	12/1/2022 10:30	A	$100
2	12/2/2022 12:15	B	$200
3	12/3/2022 15:45	C	$300

Figure 4.2 – Sales table

You are tasked with analyzing the sales data by quarter to see how different products are performing.

To extract the quarter from the `sale_date` column, you can use the following query:

```
SELECT id,
  product,
      EXTRACT(QUARTER FROM sale_date) as quarter,
      SUM(amount) as total_sales
```

```
FROM sales
GROUP BY product, quarter;
```

This query will return the following result:

id	product	quarter	total_sales
1	A	4	$100
2	B	4	$200
3	C	4	$300

Figure 4.3 – Product table

As you can see, the EXTRACT function is used to extract the quarter from the sale_date column, and the results are grouped by product and quarter. The total sales for each product during each quarter are also calculated using the SUM function.

From the result, you can see that all the sales are in the fourth quarter. With this information, you can identify which products are performing well in which quarter. Based on that, you can also plan your stock and promotions accordingly.

> **Note**
>
> It should be noted that the EXTRACT function allows you to extract only one part at a time. If you want to extract multiple parts together, you can use the DATE_TRUNC function instead.

Consider a table named flight_data with the following structure:

flight_id	departure_time	arrival_time	flight_delay
1	30:00.0	30:00.0	0
2	15:00.0	30:00.0	60
3	45:00.0	15:00.0	90

Figure 4.4 – Flight table

As an industry example, you are tasked with analyzing the flight data to see how many flights are delayed on each day of the week and how long the delays are.

To extract the day of the week from the departure_time column, you can use the following query:

```
SELECT EXTRACT(DOW FROM departure_time) as day_of_week,
       COUNT(CASE WHEN flight_delay > 0 THEN 1 ELSE NULL END) as
delayed_flights,
```

```
      AVG(flight_delay) as avg_delay
FROM flight_data
GROUP BY day_of_week;
```

This query will return the following result:

day_of_week	delayed_flight	avg_delay
2	2	75
4	1	90

Figure 4.5 – Average delay

As you can see, the EXTRACT function is used to extract the day of the week from the departure_time column, where the DOW argument specifies the day of the week. The query uses a CASE statement to check whether flight_delay is greater than 0 – if it is, then it's a delayed flight. The results are grouped by the day of the week, and it counts the delayed flights and calculates the average delay.

From the result, you can see that on day **2** (Tuesday), two flights are delayed, and the average delay is **75** minutes. On day **4** (Thursday), only one flight is delayed, and the delay is **90** minutes. This information could be useful for the airline company to find out which day of the week has more flight delays and take the necessary action accordingly to improve the flight schedule or maintenance process.

In summary, the SQL EXTRACT function allows you to organize, analyze, and simplify your data in a more efficient and meaningful way, which makes it a powerful tool in your data wrangling and analysis toolkit.

DATEDIFF()

The DATEDIFF() function is used to calculate the number of days between two dates. The DATEDIFF() function takes two arguments: a start date and an end date. The function returns the number of days between the start date and the end date. The date format must be YYYY-MM-DD or YYYYMMDD, and it must be recognized by the database system.

Here's the syntax:

```
DATEDIFF(date1, date2)
```

Please note that the first argument is the most recent date, and the second argument is the older date.

Here's an example:

```
SELECT DATEDIFF("2022-12-06", "2022-11-24") as 'Difference in days';
```

The result would be 12.

This query returns the number of days between December 6, 2022, and November 24, 2022, which is 12. The DATEDIFF() function can also take a third argument, which specifies the unit of time to use in the calculation. By default, the unit is days, but it can also be specified as hours, minutes, or seconds.

For example, you could calculate the number of hours between two dates:

```
SELECT DATEDIFF(hour, '2022-12-05 12:00:00','2022-12-06 08:00:00') AS
'Difference in hours';
```

The result would be 20.

This query will return the number of hours between December 5, 2022 at 12 A.M. and December 6, 2022 at 8 A.M., which is 20.

Case scenario

Let's say the HR team of a company has to track the tenure of its employees. The company may have a table that contains employee information, including the hire date of each employee.

In this scenario, we can use the DATEDIFF() function to calculate the number of days between the hire date and the current date. Then, using a CASE statement, we can assign the tenure of the employee depending on the number of days.

Here's an example of how this might be done:

```
SELECT
    Id,
    name as Employee_Name,
    hire_date as Hire_Date ,
    DATEDIFF(NOW(), hire_date) as Tenure,
    CASE
        WHEN DATEDIFF(NOW(), hire_date) < 365 THEN 'Less than a year'
        WHEN DATEDIFF(NOW(), hire_date) < 730 THEN 'Between 1 and 2
years'
        WHEN DATEDIFF(NOW(), hire_date) < 1095 THEN 'Between 2 and 3
years'
        ELSE 'Over 3 years'
    END as Tenure_Bucket
FROM employees
```

This query will give us the name, hire date, and tenure of each employee, and it will also assign a tenure bucket to them depending on the number of days. It will first calculate the number of days from the hire date of each employee to the current date and store it in the Tenure column. Then, using the CASE statement, it will decide the tenure bucket the employee falls into, as follows.

id	Employee_Name	Hire_Date	Tenure	Tenure_Bucket
1	Alex Johnson	1/1/2020	1109	3 years
2	Emily Davis	5/1/2019	1354	4 years
3	Michael Thompson	12/1/2018	1505	5 years
4	Sarah Rodriguez	2/1/2019	1443	6 years

Figure 4.6 – Employee table

Let's say we have a retail company, Max, where the company wants to analyze the behavior of its customers. The company may have a table that contains customer purchase information, including the date and time of each purchase. By using the DATEDIFF function, the company could calculate the number of days between a customer's first purchase and their most recent purchase. This information could then be used to segment customers into different groups based on their level of engagement or loyalty to the company. For example, customers who have made a purchase within the last 30 days could be considered "active" customers, while those who have not made a purchase in 60 days or more could be considered "inactive." With this information, the company could then target marketing and loyalty efforts to different segments of customers in a more effective way.

TIMEDIFF()

The TIMEDIFF() function in SQL is used to calculate the difference between two times or DateTime expressions. The function returns the difference as a time value.

Please note that the first argument is the most recent DateTime and the second argument is the older DateTime.

Here's the syntax:

```
TIMEDIFF(time1, time2)
```

Here's an example:

```
SELECT TIMEDIFF("2022-12-11 16:20:45", "2022-12-11 12:20:50");
```

The result would be 03:59:55.

This gives a difference of around 4 hours between the two timeframes.

Case scenario

Let's say we have a table called employees that has the following columns: id, name, clock_in, and clock_out. If you wanted to find the amount of time each employee worked, you could use the following query:

```
SELECT id, name, clock_in, clock_out,
       TIMEDIFF(clock_out, clock_in) as hours_worked
FROM employees;
```

This query will return a new column called `hours_worked`, which contains the difference in time between each row's `clock_out` and `clock_in` times for each employee.

id	name	clock_in	clock_out	hours_worked
1	Alex Johnson	9:00:00	17:00:00	8
2	Emily Davis	8:30:00	16:30:00	8
3	Michael Thompson	9:15:00	18:00:00	8:45
4	Sarah Rodriguez	7:45:00	15:45:00	8

Figure 4.7 – Employees table

One interesting industry example for using the `TIMEDIFF()` function in SQL could be in the logistics industry, specifically for calculating the arrival and departure times of trucks.

Let's say that a logistics company has a database with a table called `schedules` that contains information about the arrival and departure times of trucks at different stations. The table has the following columns: `id`, `truck_number`, `route`, `departure_time`, and `arrival_time`.

The company wants to know how long each truck takes to travel between each pair of stations. They could use the `TIMEDIFF()` function to calculate the difference between the arrival time and departure time at each station:

```
SELECT id, truck_number, route, departure_time, arrival_time,
       TIMEDIFF(arrival_time, departure_time) as travel_time
FROM schedules;
```

This query would return a new column called `travel_time`, which contains the difference in time between each row's `arrival_time` and `departure_time` for each truck.

id	truck_number	route_id	departure_time	arrival_time	travel_time
1	TK-123	R-1	8:00 AM	12:00 PM	4:00
2	TK-456	R-2	9:00 AM	3:00 PM	6:00
3	TK-789	R-3	10:00 AM	5:00 PM	7:00
4	TK-321	R-4	11:00 AM	9:00 PM	10:00

Figure 4.8 – Truck table

This information can be useful for the company in several ways:

- To optimize schedules by identifying the routes that take longer than expected and looking for ways to shorten the travel time
- To identify delays by comparing the planned travel time with the actual travel time
- To predict and plan maintenance schedules for transportation vehicles

The company could also use the travel time to provide more accurate arrival time predictions for customers who use the company's logistic services.

Please note that the order of arguments in `TIMEDIFF(time2, time1)` matters since it calculates `time2` minus `time1` and it will return a negative value if `time2` comes before `time1`. The returned result is in the HH:MM:SS format. If you want to have the result in hours or minutes, you need to do a conversion based on the returned result.

The `DATEDIFF/TIMEDIFF` functions are very useful in the following scenarios:

- To find the difference between two date values or two time values. Let's say we want to find the difference between an order date and a shipping date. The DATEDIFF function will help us to find the number of days between them and the `TIMEDIFF()` function will be used if we want to analyze the time gap between two deliveries.
- To calculate age in years with respect to the current date.
- To calculate the duration of an event.
- To identify how long it takes for an employee to complete a task.

DATE_ADD()

The `DATE_ADD()` function in SQL is used to add a specified date/time interval to a given date or DateTime value. The resulting date or datetime value is then returned.

Here's the syntax:

```
DATE_ADD(date_value, INTERVAL date_value/time_value add_unit)
```

The `add_unit` can be any of the following units that can be added to the date:

`MICROSECOND, SECOND, MINUTE, HOUR, DAY, WEEK, MONTH, QUARTER, YEAR`, and so on.

Case scenario

Let's say you have a table called `orders` that has the following columns: `id`, `order_date`, and `delivery_date`. If you wanted to find the delivery date for each order that is three days after the order date, you could use the following query:

```
SELECT id, order_date, delivery_date
FROM orders
WHERE delivery_date = DATE_ADD(order_date, INTERVAL 3 DAY);
```

This query will return a new column called `delivery_date`, which is three days after the order date.

id	order_date	delivery_date
1	1/1/2022	1/4/2022
2	12/1/2021	12/4/2021
3	11/1/2021	11/4/2021
4	10/1/2021	10/4/2021

Figure 4.9 – Order table

`DATE_ADD()` can also be used to subtract time intervals from a given date by using a negative value for the interval expression:

```
SELECT DATE_ADD(date_column, INTERVAL -value interval_unit) AS new_
date
FROM your_table;
```

Suppose you have a table named `loans` that contains information about loans issued by a bank, such as the loan amount, interest rate, start date, and end date. You want to write a query that will identify loans that are about to reach the end of their term in the next 30 days.

Here is how you could use the `DATE_ADD()` function to achieve this:

```
SELECT loan_id, loan_amount, interest_rate, start_date, end_date
FROM loans
WHERE end_date BETWEEN CURDATE() AND DATE_ADD(CURDATE(), INTERVAL 30
DAY)
```

This query will select all `loan_id`, `loan_amount`, `interest_rate`, `start_date`, and `end_date` of loans that are going to end in the next 30 days. This information is useful to remind the bank to prepare for loan repayment or to reach out to customers to renew their loans.

loan_id	loan_amount	interest_rate	start_date	end_date
1	$10,000	5%	1/1/2022	1/1/2023
2	$15,000	4.50%	12/1/2021	12/28/2022
3	$20,000	6%	11/1/2021	12/24/2022
4	$25,000	5.50%	10/1/2021	12/20/2022

Figure 4.10 – Loans table

Let's say the bank is also interested in following additional metrics that can also be useful in real-time analytics:

- Details of loans that are going to end in the next 30 minutes
- Details of loans that are going to end in the next 2 hours
- Details of loans that are going to end in the next 6 months
- Details of loans that are going to end in the next 2 quarters
- Details of loans that are going to end in the next 3 years

We can generate these metrics with the following SQL queries using the date_add function.

Here's the code for details of loans that are going to end in the next 30 minutes:

```
SELECT loan_id, loan_amount, interest_rate, start_date, end_date
FROM loans
WHERE end_date BETWEEN CURDATE() AND DATE_ADD(CURDATE(), INTERVAL 30
MINUTE)
```

Here's the code for details of loans that are going to end in the next 2 hours:

```
SELECT loan_id, loan_amount, interest_rate, start_date, end_date
FROM loans
WHERE end_date BETWEEN CURDATE() AND DATE_ADD(CURDATE(), INTERVAL 2
HOUR)
```

Here's the code for details of loans that are going to end in the next 6 months:

```
SELECT loan_id, loan_amount, interest_rate, start_date, end_date
FROM loans
WHERE end_date BETWEEN CURDATE() AND DATE_ADD(CURDATE(), INTERVAL 6
MONTH)
```

Here's the code for details of loans that are going to end in the next 2 quarters:

```
SELECT loan_id, loan_amount, interest_rate, start_date, end_date
FROM loans
WHERE end_date BETWEEN CURDATE() AND DATE_ADD(CURDATE(), INTERVAL 2
QUARTER)
```

Here's the code for details of loans that are going to end in the next 3 years:

```
SELECT loan_id, loan_amount, interest_rate, start_date, end_date
FROM loans
WHERE end_date BETWEEN CURDATE() AND DATE_ADD(CURDATE(), INTERVAL 3
YEAR)
```

In a financial setting, it is important to keep track of when loans are set to expire so that necessary actions can be taken to ensure that payments are made on time and to avoid any potential defaults or late fees. Additionally, this information could also be used to forecast future cash flow and to plan future lending strategies.

An interesting industry example for using the DATE_ADD function in SQL could be in the retail industry, specifically for managing inventory and planning to restock.

Let's say that a retail company has a database with a table called products that contains information about the products they sell, including the product name, current stock level, and the last restocked date. The table has the following columns: id, name, stock, and last_restocked.

The company wants to be able to plan its restocking schedule and predict when products will run low on stock so that it can reorder them in a timely manner. It could use the DATE_ADD function to calculate when it expects the current stock to run out based on the average daily usage of each product:

```
SELECT id, name, stock, last_restocked_date, average_daily_usage
       DATE_ADD(last_restocked, INTERVAL stock/average_daily_usage
DAY) as expected_stock_out
FROM products;
```

This query will return a new column called expected_stock_out, which contains the date when the current stock of each product is expected to run out.

id	name	stock	last_restocked_date	average_daily_usage	expected_stock_out
1	TV	100	1/1/2022	2	2/20/2022
2	Headphones	150	12/1/2021	5	12/31/2021
3	Laptop	200	11/1/2021	7	11/29/2021
4	Power bank	250	10/1/2021	3	12/23/2021

Figure 4.11 – Products table

This information can be useful to the company in several ways:

- To schedule the restocking of products in a timely manner
- To identify which products are running low on stock and need to be reordered sooner
- To predict sales trends and help in forecasting

The company can also use this information to create a restocking schedule for each product and set alerts for when a product is running low on stock so that they can reorder it in a timely manner.

DATE_SUB()

The DATE_SUB() function works similarly to date_add. The only difference is it subtracts the date/time interval from a date and returns a date.

Here's the syntax:

```
DATE_SUB(date_value, INTERVAL date_value/time_value sub_unit)
```

sub_unit can be any of the following units that can be subtracted from the date.

MICROSECOND, SECOND, MINUTE, HOUR, DAY, WEEK, MONTH, QUARTER, YEAR, and so on.

Case scenarios

Let's say the company wants to look at customer acquisition trends over a period of time. The company could use DATE_SUB() to count the number of new customers acquired over the last 30 days.

id	customer_name	signup_date	current_date
1	Emily Davis	12/8/2022	12/30/2022
2	Michael Thompson	12/15/2022	12/30/2022
3	Sarah Rodriguez	11/15/2022	12/30/2022
4	David Kim	10/31/2022	12/30/2022

Figure 4.12 – Customer signup table

Lets write a code to generate the trends:

```
SELECT COUNT(*)as count_customers FROM customer_data
WHERE signup_date >= DATE_SUB(CURDATE(), INTERVAL 30 DAY);
```

count_customers
2

Figure 4.13 – Customer count

This query would retrieve the number of new customers who have signed up in the last 30 days.

Next, let's say there is a retail company that wants to analyze sales data for the past seven days. In this way, the retail company can check sales trends over a period of time:

```
SELECT
  product_id,
  product_name,
```

```
  SUM(total_sales) as total_sales
FROM sales_data
WHERE date >= DATE_SUB(CURDATE(), INTERVAL 7 DAY)
Group By product_id, product_name;
```

This query would retrieve the sum of all sales made in the last seven days.

product_id	product_name	total_sales
1	TV	$20,000
2	Headphones	$5,000
3	Laptop	$50,000
4	Power Bank	$7,000

Figure 4.14 – Product table with sales data

The company also wants to check the following metrics, which are very useful for real-time analytics:

- Total sales that took place in the last 40 minutes
- Total sales that took place in the last 2 hours
- Total sales in the last 7 months
- Total sales in the last 2 quarters
- Total sales in the last 3 years

We can generate these metrics with the following SQL queries using the DATE_SUB() function.

Here's the query for the total sales that took place in the last 40 minutes:

```
SELECT SUM(total_sales) FROM sales_data
WHERE date >= DATE_SUB(CURDATE(), INTERVAL 40 MINUTE);
```

Here's the query for the total sales that took place in the last 2 hours:

```
SELECT SUM(total_sales) FROM sales_data
WHERE date >= DATE_SUB(CURDATE(), INTERVAL 2 HOUR);
```

Here's the query for the total sales in the last 7 months:

```
SELECT SUM(total_sales) FROM sales_data
WHERE date >= DATE_SUB(CURDATE(), INTERVAL 7 MONTH);
```

Here's the query for the total sales in the last 2 quarters:

```
SELECT SUM(total_sales) FROM sales_data
WHERE date >= DATE_SUB(CURDATE(), INTERVAL 2 QUARTER);
```

Here's the query for the total sales in the last 3 years:

```
SELECT SUM(total_sales) FROM sales_data
WHERE date >= DATE_SUB(CURDATE(), INTERVAL 3 YEAR);
```

An interesting example of using DATE_SUB in SQL in a real-world scenario could be in the field of healthcare.

Imagine a hospital that wants to track the number of patients who were discharged in the last 30 days and were readmitted within the next 30 days. They want to track this data to identify any common reasons for readmission and improve their discharge process.

To accomplish this, the hospital would have two relevant tables in its database: one for patient admissions and one for patient discharges. The following SQL query could be used to retrieve the necessary data:

```
SELECT COUNT(*) FROM admissions
WHERE discharge_date >= DATE_SUB(CURDATE(), INTERVAL 30 DAY)
AND admission_date <= DATE_SUB(CURDATE(), INTERVAL 30 DAY);
```

This query would return the number of patients who were discharged in the last 30 days and were readmitted within the next 30 days.

The hospital could also use this data to identify the common reasons for readmission by joining the admission table with other tables, such as diagnosis and treatment tables.

In this scenario, DATE_SUB is used to specify the date range for the query, making it easy to track the number of patients who were discharged in the last 30 days and were readmitted within the next 30 days. This can help the hospital to improve the discharge process and reduce the number of readmissions.

Here are some practical uses of the DATE_SUB and DATE_ADD functions:

- **Time-based analytics**: The DATE_SUB function is used a lot to analyze how data was trending in the past (day, month, quarter, and year). This helps in forecasting data and in building predictive data models.

- **Data-driven insights**: Can be used to generate new trends and insights based on historical data.

- **Dynamic reporting**: Can be used to create dynamic reports that automatically update based on the current date.

- **Improved performance**: By using DATE_SUB/DATE_ADD to filter data based on a specific date range, you can improve the performance of your queries by reducing the amount of data that needs to be processed.

- **Simplified queries**: DATE_SUB/DATE_ADD can make your SQL queries more readable and easier to understand by eliminating the need for complex date calculations in the WHERE clause.

- **Consistency**: DATE_SUB/DATE_ADD can help to ensure consistency in your data analysis by providing a standardized way to specify date ranges across different queries and reports.

- **Flexibility**: DATE_SUB/DATE_ADD allows you to subtract different time units, such as days, months, and years, making it a flexible tool for working with dates in SQL.

DATE_FORMAT()

In SQL, the DATE_FORMAT() function is used to format a date value according to a specified format. The function takes two arguments: the date value to be formatted and a format string that specifies how the date should be represented.

The format string can contain a variety of placeholders, such as %Y for the year, %m for the month, and %d for the day. For example, the following query selects the current date and formats it as yyyy-mm-dd.

Here's the syntax:

```
DATE_FORMAT(date, format)
```

DATE – The date entered, which needs to be formatted.

Format – The specific format into which the date entered has to be formatted.

Case scenarios

Scenario 1

A cybersecurity company needs to analyze log data from a firewall, but the timestamp field is in the yyyy-mm-dd hh:mm:ss format. In order to create a report of the number of events that occurred each day, the company needs to format the timestamp field to show only the date:

```
SELECT DATE_FORMAT(timestamp, '%Y-%m-%d') AS date, COUNT(*) as events
FROM firewall_logs
GROUP BY date
ORDER BY date;
```

In this example, the DATE_FORMAT() function is used to format the timestamp field to only show the date using the format specifier '%Y-%m-%d', the data is grouped by date, and the number of events is counted.

Scenario 2

Let's say a cybersecurity company needs to analyze log data from a web server, but the `timestamp` field is in the format dd-mm-yyyy hh:mm:ss. In order to create a report of the number of events that occurred each hour, the company needs to format the `timestamp` field to show only the hour:

```
SELECT DATE_FORMAT(timestamp, '%H') AS hour, COUNT(*) as events
FROM web_server_logs
GROUP BY hour
ORDER BY hour;
```

In this example, the `DATE_FORMAT()` function is used to format the `timestamp` field to only show the hour using the format specifier '%H', the data is grouped by hour, and the number of events is counted.

Scenario 3

A cybersecurity company needs to analyze log data from an intrusion detection system, but the `timestamp` field is in the format dd/mm/yyyy hh:mm:ss. In order to create a report of the number of events that occurred each month, the company needs to format the `timestamp` field to show only the month:

```
SELECT DATE_FORMAT(timestamp, '%m') AS month, COUNT(*) as events
FROM intrusion_detection_logs
GROUP BY month
ORDER BY month;
```

In this example, the `DATE_FORMAT()` function is used to format the `timestamp` field to only show the month using the % format specifier.

timestamp	event_type	event_description	event_outcome	user_name	source_ip
5/10/21 12:00 PM	login_attempt	Failed login attempt from user `admin` from IP	Failure	admin	192.168.1.100
15/1/2021 12:05:01 PM	login_attempt	Successful login for user `jdoe` from IP	Success	jdoe	192.168.1.101
5/7/21 12:10 PM	file_access	Access to `secret.txt` file by user `jdoe`	Success	jdoe	192.168.1.101

Figure 4.15 – intrusion_detection_logs

Date	Hour	Month
5/10/2021	12	10
15/1/2021	12	1
5/7/2021	12	7

Figure 4.16 – Results of the preceding three queries

Suppose you have a table named `sales` that contains information about the sales made by a company, including the date of the sale and the total revenue of each sale. The date values are stored as the DATE data type, but you want to analyze the data by the financial quarter of the year.

sale_id	sale_date	Time	revenue
1	1/1/2023	10:00 AM	$2,000
2	1/1/2023	11:00 AM	$5,000
3	1/1/2023	12:00 PM	$3,500
4	1/1/2023	2:00 PM	$6,000
5	1/2/2023	9:00 AM	$8,500

Figure 4.17 – Results of the preceding three queries

You could use the following SQL query to format the date values in the `sales` table as a financial quarter and then use the formatted values to group the sales and calculate the total revenue for each quarter:

```
SELECT
    DATE_FORMAT(sale_date, '%Y-Q%q') as sale_quarter,
    SUM(revenue) as total_revenue
FROM sales
GROUP BY sale_quarter;
```

This query first formats the `sale_date` values in the `'%Y-Q%q'` format, where Q stands for quarter, `%q` is the quarter of the year as a decimal number (1, 2, 3, 4), and the year is in the `%Y` format. Next, the query groups the sales by quarter and calculates the total revenue for each quarter. The result is a table showing the total revenue generated by the company for each financial quarter.

This can be useful for business owners to compare the performance of the company over the quarters and make an informed decision on budgeting, forecasting, and inventory management.

In addition, you could use this function to extract other information too, such as the date of a day of the week or month of the year, by formatting the date in a specific format. This is just one example of how the DATE_FORMAT() function can be used to analyze data in a business context, and you can use the function to format dates in various formats, extract different information, and further analyze it to extract meaningful insights.

STR_TO_DATE()

Similar to the DATE_FORMAT() function, there is another function in SQL called STR_TO_DATE() that has similar functionality.

The STR_TO_DATE() function is used to convert a string value into a date value. This function takes two arguments: the string value to be converted and a format string that specifies the format of the input string.

Here's the syntax:

```
STR_TO_DATE(string, format)
```

string means the value is in string format.

format is the specific format in which the date entered has to be formatted.

> **Note**
> Both DATE_FORMAT and STR_TO_DATE work the same way. You can use either one of them.

The format string can contain a variety of placeholders, such as %Y for the year, %m for the month, and %d for the day.

The STR_TO_DATE() function in SQL is useful in the retail industry when working with data that has been imported from external sources and may be in different date formats.

Here are a few examples of how the STR_TO_DATE() function could be used in retail industry scenarios.

Case scenarios

Let's explore a few real-life case scenarios that illustrate the practical applications of the concepts we've discussed.

Scenario 1

A retail company receives a CSV file of customer data, but the date_of_ birth field is in different formats, such as dd-mm-yyyy, mm/dd/yyyy, yyyy-mm-dd, and so on. To analyze the customer data effectively, the company needs to convert the date_of_birth field into a consistent format:

```
SELECT
    customer_name,
    STR_TO_DATE(date_of_birth, '%d-%m-%Y') as date_of_birth_new,
    age,
    city
FROM customers
```

In this example, the STR_TO_DATE() function is used to convert the date_of_birth field from a string into a date value using the format specifier '%d-%m-%Y'.

customer_name	date_of_birth
John Smith	5/12/1980
Emily Johnson	1/23/1990
Michael Brown	3/15/1975
Olivia Garcia	1/7/2000
David Martinez	9/20/1985

customer_name	date_of_birth_new
John Smith	12/5/1980
Emily Johnson	23/1/1990
Michael Brown	15/3/1975
Olivia Garcia	1/7/2000
David Martinez	20/9/1985

Figure 4.18 – Date conversion using STR_TO_DATE()

Scenario 2

A retail company receives data from multiple sources on purchase dates, but the dates are in different formats, such as yyyy-mm-dd, dd/mm/yyyy, and so on. To analyze the purchase data effectively, the company needs to convert the dates into a consistent format:

```
SELECT
    order_id,
    STR_TO_DATE(order_date, '%Y-%m-%d') as order_date,
    product_name,
    quantity
FROM orders
```

In this example, the STR_TO_DATE() function is used to convert the order_date field from a string into a date value using the format specifier '%Y-%m-%d'.

Scenario 3

A retail company receives data from multiple sources on return dates, but the dates are in different formats, such as dd-mm-yyyy hh:mm:ss, mm/dd/yyyy hh:mm:ss, and so on. In order to analyze the return data effectively, the company needs to convert the dates into a consistent format:

```
SELECT
    return_id,
    STR_TO_DATE(return_date, '%d-%m-%Y %H:%i:%s') as return_date,
    product_name,
    reason
FROM returns
```

In this example, the STR_TO_DATE() function is used to convert the return_date field from a string into a date value using the format specifier '%d-%m-%Y %H:%i:%s', which includes both the date and time.

order _id	order _date	product _name	return _date		order _id	order _date	product _name	return _date
1001	5/1/2021	iPhone 12			1001	5/1/2021	iPhone 12	
1002	2/5/2021	MacBook Pro	5/20/2021		1002	2/5/2021	MacBook Pro	5/20/21 12:00 AM
1003	2021-03-05	Samsung TV			1003	3/5/2021	Samsung TV	
1004	05/04/21	AirPods Pro			1004	5/4/2021	AirPods Pro	
1005	5/5/2021	Nintendo Switch	2021-05-07		1005	5/5/2021	Nintendo Switch	5/7/21 12:00 AM

Figure 4.19 – Date modification based on queries 2 and 3

These are just a few examples of how the STR_TO_DATE() function can be used in a retail industry scenario to convert data from different date formats into a consistent format for analysis and reporting. This can help the company to gain insights and make better business decisions.

The main keywords are the following:

- %a – This abbreviation refers to the name of the weekday. It ranges from Sunday to Saturday.
- %b – This abbreviation refers to the month's name. It ranges from January to December.
- %M – This abbreviation refers to the full month's name from January to December.
- %m – This abbreviation refers to the month number from 01 to 12.
- %c – This abbreviation refers to the number of the month. It ranges from 0 to 12.
- %e – This abbreviation refers to the day of the month as a numeric value. It ranges from 0 to 31.
- %H – This abbreviation refers to an hour. It ranges from 00 to 23.
- %i – This abbreviation refers to minutes. It ranges from 00 to 59.
- %j – This abbreviation refers to the day of the year. It ranges from 001 to 366.
- %p – This abbreviation refers to AM or PM.
- %S – This abbreviation refers to seconds. It ranges from 00 to 59.
- %W – This abbreviation refers to the weekday name from Sunday to Saturday.
- %w – This abbreviation refers to the weekday number from 0 to 6. 0 being Sunday and 6 being Saturday.
- %Y – This abbreviation refers to the year as a numeric value of 4 digits.
- %y – This abbreviation refers to the year as a numeric value of 2 digits.

Scenario 4

Suppose you have a table named `sales` that contains information about the sales made by a company, including the date of the sale and the total revenue of each sale. The date values are stored as strings in the format mm/dd/yyyy.

sale_id	sale_date	time	revenue
1	1/1/2023	10:00 AM	$2,000
2	1/1/2023	11:00 AM	$5,000
3	1/1/2023	12:00 PM	$3,500
4	1/1/2023	2:00 PM	$6,000
5	1/2/2023	9:00 AM	$8,500

Figure 4.20 – Sales table

You want to use the data in this table to analyze the sales performance of the company over time. Specifically, you want to determine the total revenue generated by the company each month, so you can track whether revenue is increasing or decreasing over time.

You could use the following SQL query to convert the date strings into the proper DATE format and then use the DATE values to group the sales by month and calculate the total revenue for each month:

```
SELECT
    MONTH(STR_TO_DATE(sale_date, '%m/%d/%Y')) as sale_month,
    SUM(revenue) as total_revenue
FROM sales
GROUP BY sale_month;
```

This query first converts the `sale_date` strings into the DATE format using the STR_TO_DATE() function and then extracts the month from each DATE value using the MONTH() function. Next, the query groups the sales by month and calculates the total revenue for each month. The result is a table showing the total revenue generated by the company for each month.

You could also use the same data to analyze the sales performance based on the day of the week or year-wise.

This will be useful information to business owners as it could give them an idea of the most and least profitable time period in their business, which could be useful when they plan their sales strategy, inventory management, and budgeting.

Extracting the current date and time

Comparing the data populated with current data helps in understanding how data is trending as of that date compared to historical data.

The following functions help in extracting the current date:

- CURDATE() or CURRENT_DATE()
- CURTIME() or CURRENT_TIME()
- CURRENT_TIMESTAMP()
- NOW()
- SYSDATE()
- CAST()

CURDATE() / CURRENT_DATE()

In SQL, the CURDATE() or CURRENT_DATE() function is used to return the current date. The function takes no arguments and returns the date in the format YYYY-MM-DD.

Here's the syntax:

```
CURDATE() or CURRENT_DATE()
```

Here's an example:

```
SELECT CURDATE();
```

This query will return the current date.

You can also use this function to compare with a date or timestamp field in a table like this:

```
SELECT * FROM table_name WHERE date_field <= CURDATE();
```

This will select all records from table_name where date_field is less than or equal to the current date.

Case scenario 1

In the e-commerce industry, the CURDATE() or CURRENT_DATE() function can be used in a variety of ways to analyze sales data and make business decisions. Here's an example of a scenario in which this function might be used:

date_placed	product_id	product_name	units	price
5/1/2021	1211	iPhone 12	20	2000
5/2/2021	1213	iPhone 13	15	1500
5/3/2021	1234	iPhone 14	25	2500
5/4/2021	1234	iPhone 14	30	3000
5/5/2021	1234	iPhone 14	25	2500

Figure 4.21 – Orders table

A company wants to track the daily sales of a particular product. It has a database with a table called orders that contains information about each order, including the date the order was placed, the product ID, and the price of the product.

The company wants to retrieve the total sales of a particular product ID for the current date. It would use the CURDATE() function in a SQL query to retrieve the current date and then use that date in a WHERE clause to filter the orders table for only the orders placed on the current date.

Here's an example of a SQL query that would accomplish this:

```
SELECT SUM(price) FROM orders WHERE date_placed = CURDATE() AND
product_id = 1234;
```

This query would retrieve the total sales of product ID 1234 for the current date by summing up the price of all the rows in the orders table where the date_placed column is equal to the current date and the product_id column is equal to 1234. The CURDATE() function is used in the WHERE clause to filter the orders table for only the rows where the date_placed column is equal to the current date. Assuming the current date is 5/3/2021, we get the following result.

sum(price)
2500

Figure 4.22 – Result table

Alternatively, you can also use the query with a GROUP BY clause to get the sales of a product for a particular date range:

```
SELECT date_placed, SUM(price) FROM orders WHERE product_id = 1234
GROUP BY date_placed;
```

This query would retrieve the total sales of product ID 1234 for each date in the date_placed column by summing up the prices of all the rows in the orders table where the product_id column is equal to 1234 and grouping them by the date_placed column.

date_placed	sum(price)
5/3/2021	2500
5/4/2021	3000
5/5/2021	2500

Figure 4.23 – Result table

This kind of analysis can help the company make decisions about which products to promote, when to run sales, and how much inventory to stock. Other use cases may include tracking the number of unique customers that made a purchase on the current date or analyzing the average order value for the current date.

Case scenario 2

In the finance industry, the CURDATE() or CURRENT_DATE() function can be used in a variety of ways to retrieve information from the database. Here's a scenario:

Date	customer_name	account_number	transaction_type	amount	description
5/1/2021	Sam Smith	1234567890	deposit	$1,000.00	Salary credit from XYZ company
5/1/2021	Emily Johnson	1234567891	withdrawal	$ 500.00	Cash withdrawal at ATM
5/2/2021	Michael Brown	1234567892	transfer	$ 250.00	Transfer to savings account
5/3/2021	Olivia Garcia	1234567893	bill payment	$ 100.00	Electric bill payment to ABC utility company
5/4/2021	David Martinez	1234567894	deposit	$2,000.00	Direct deposit from DEF company

Figure 4.24 – Transactions table

A bank wants to track the daily transactions made by its customers. It has a database with a table called transactions that contains information about each transaction, including the date the transaction was made, the account number, the type of transaction, and the amount.

The bank wants to retrieve the total amount of transactions made on the current date. It would use the CURDATE() function in a SQL query to retrieve the current date and then use that date in a WHERE clause to filter the transactions table for only the transactions made on the current date.

Here's an example of a SQL query that would accomplish this:

```
SELECT SUM(amount) FROM transactions WHERE date = CURDATE();
```

This query would retrieve the sum of the total amount of all the transactions made on the current date. The CURDATE() function is used in the WHERE clause to filter the transactions table for only the rows where the date column is equal to the current date.

Additionally, the bank may also want to know the total number of transactions made by a particular account on the current date:

```
SELECT COUNT(*) FROM transactions WHERE date = CURDATE() and account_
number = '1234567890';
```

This query would retrieve the total number of transactions made by account number '1234567890' on the current date. The CURDATE() function is used in the WHERE clause to filter the transactions table for only the rows where the date column is equal to the current date. The query also filters the transactions for a particular account number.

In the finance industry, these are just a few examples of how the CURDATE() or CURRENT_DATE() function can be used to retrieve relevant information from a database. Other use cases may include tracking the number of transactions made by different types of accounts (e.g., savings, checking, etc.) on the current date or analyzing the total amount of transactions made by a particular branch on the current date.

CURTIME() / CURRENT_TIME()

The CURTIME() or CURRENT_TIME() function is used to retrieve the current time. The function returns the current time in the HH:MM:SS format.

Here's the syntax:

```
CURTIME();
```

Here's an example:

```
SELECT CURTIME();
```

This statement would return the current time in the format HH:MM:SS. Alternatively, you can use the NOW() function, which returns the current date and time. You can use this function in various scenarios, such as tracking the time of login, time of transaction, and so on.

For example, a company wants to track the time of login of their employees. It has a database with a table called `employee_logins` that contains information about each login, including the date and time of login, the employee ID, and the department.

employee_id	employee_name	login_time	logout_time	department
1	Sam Sai	5/1/2021 9:00	5/1/2021 17:00	HR
2	Emily Beth	5/2/2021 8:30	5/2/2021 17:00	Operations
3	Michael Joe	5/3/2021 9:00	5/3/2021 17:00	Operations
4	Erine Wan	5/4/2021 8:00	5/4/2021 17:00	Service

Figure 4.25 – Employee logins table

The company wants to retrieve the time of the last login of an employee. It would use the `CURTIME()` function in a SQL query to retrieve the current time and then use that time in a `WHERE` clause to filter the `employee_logins` table to only the logins made at the current time:

```
SELECT employee_id, department FROM employee_logins WHERE time_of_
login = CURTIME();
```

This query would retrieve the employee ID and department of all the employees who logged in at the current time. The `CURTIME()` function is used in the `WHERE` clause to filter the `employee_logins` table for only the rows where the `time_of_login` column is equal to the current time.

Case scenario 1

A transportation company wants to track the times of departure and arrival of their buses. They have a database with a table called `bus_schedule` that contains information about each bus, including the departure and arrival times, the route, and the bus number.

route	bus_number	departure_location	arrival_location	departure_time	arrival_time
Route 1	78570	Downtown	Airport	8:00 AM	9:30 AM
Route 2	38572	Airport	Downtown	4:00 PM	5:30 PM
Route 3	28563	Central Station	University	6:00 AM	7:15 AM
Route 4	87321	University	Central Station	3:00 PM	4:15 PM

Figure 4.26 – Bus schedule table

The company wants to retrieve the buses arriving at the current time. It would use the `CURTIME()` function in a SQL query to retrieve the current time and then use that time in a `WHERE` clause to filter the `bus_schedule` table to only the buses arriving at the current time.

Here's an example of a SQL query that would accomplish this:

```
SELECT bus_number, route FROM bus_schedule WHERE arrival_time =
CURTIME();
```

This query would retrieve the bus number and route of all the buses that are arriving at the current time. The CURTIME() function is used in the WHERE clause to filter the bus_schedule table to only the rows where the arrival_time column is equal to the current time.

Additionally, the company may also want to know the buses that are scheduled to depart in the next hour:

```
SELECT bus_number, route FROM bus_schedule WHERE departure_time >=
CURTIME() AND departure_time < DATE_ADD(CURTIME(), INTERVAL 1 HOUR);
```

This query would retrieve the bus number and route of all the buses that are scheduled to depart in the next hour. The CURTIME() function is used in the WHERE clause along with the DATE_ADD function to filter the bus_schedule table to only the rows where the departure_time column is greater than or equal to the current time and less than the current time plus one hour.

In the transportation industry, these are just a few examples of how the CURTIME() or CURRENT_TIME() function can be used to retrieve relevant information from the database. Other use cases may include tracking the number of buses delayed or analyzing the average arrival time of the buses.

Case scenario 2

In the stock market, the CURTIME() or CURRENT_TIME() function can be used in a variety of ways to retrieve information from a database. Here's a scenario:

A financial institution wants to track the real-time stock prices of various companies. It has a database with a table called stock_prices that contains information about each stock, including the date and time of the price, the stock symbol, and the price.

company_name	stock_symbol	current_price	open_price	high_price	low_price	volume	date_time
Apple Inc.	AAPL	$132.50	$131.00	$133.00	$130.50	4,500,000	1/15/2023 9:30 AM
Apple Inc.	AAPL	$134.50	$131.00	$133.50	$130.20	4,500,100	1/15/2023 10:30 AM
Microsoft Corp	MSFT	$250.00	$248.50	$251.00	$247.50	3,200,000	1/15/2023 9:30 AM
Microsoft Corp	MSFT	$255.00	$248.50	$252.00	$249.50	3,202,000	1/15/2023 11:30 AM
Amazon.com Inc	AMZN	$3,200.00	$3,150.00	$3,250.00	$3,100.00	1,500,000	1/15/2023 9:30 AM

Figure 4.27 – Stock price table

The institution wants to retrieve the current stock prices of a particular company. It would use the CURTIME() function in a SQL query to retrieve the current time and then use that time in a WHERE clause to filter the stock_prices table to only the stock prices at the current time:

```
SELECT
stock_symbol,
current_price
FROM stock_prices
WHERE date_time = CURTIME() AND company_name=' Apple Inc.';
```

This query would retrieve the stock symbol and price of the stock of Apple Inc. at the current time. The CURTIME() function is used in the WHERE clause to filter the stock_prices table to only the rows where the date_time column is equal to the current time and the stock symbol is ' Apple Inc.'.

Additionally, the financial institution may also want to know the list of stocks that have shown a significant price increase or decrease in the last hour:

```
SELECT
stock_symbol,
current_price
FROM stock_prices
WHERE date_time = CURTIME() AND company_name=' Apple Inc.';
```

This query would retrieve the stock symbol and the price change (compared to the last hour) of the stocks that have shown a significant price increase or decrease in the last hour. The CURTIME() function is used in the WHERE clause along with the DATE_SUB function to filter the stock_prices table to only the rows where the date_time column is equal to the current time. The query also uses a subquery to calculate the price change by comparing the current price with the price from the last hour and returns the stocks that have shown a price change greater than 5 or less than -5.

In the stock market, these are just a few examples of how the CURTIME() or CURRENT_TIME() function can be used to retrieve relevant information from a database. Other use cases may include tracking the highest and lowest stock prices for the day or analyzing the average stock price for a particular company at the current time.

CURRENT_TIMESTAMP() / NOW() / SYSDATE()

In SQL, the CURRENT_TIMESTAMP(), NOW(), and SYSDATE() functions are used to retrieve the current date and time.

The CURRENT_TIMESTAMP() function returns the current date and time in the format YYYY-MM-DD HH:MM:SS. Here is an example of how you might use the CURRENT_TIMESTAMP() function in a SQL statement:

Here's the syntax:

```
CURRENT_TIMESTAMP()
```

Here's an example:

```
SELECT CURRENT_TIMESTAMP();
```

This statement would return the current date and time in the format YYYY-MM-DD HH:MM:SS.

The NOW() function also returns the current date and time in the format YYYY-MM-DD HH:MM:SS. Here is an example of how you might use the NOW() function in a SQL statement:

```
SELECT NOW();
```

This statement would return the current date and time in the format YYYY-MM-DD HH:MM:SS.

The SYSDATE() function also returns the current date and time in the format YYYY-MM-DD HH:MM:SS. Here is an example of how you might use the SYSDATE() function in a SQL statement:

```
SELECT SYSDATE();
```

This statement would return the current date and time in the format YYYY-MM-DD HH:MM:SS.

Case scenario 1

A social media company wants to track the activity on its platform. It has a database with a table called user_activity that contains information about each user's activity, including the date and time of the activity, the user's ID, and the type of activity.

user_id	date_time	activity_type
1	1/1/23 10:00 AM	Post
2	1/1/23 11:00 AM	Comment
3	1/1/23 12:00 AM	Share
4	1/1/23 2:00 PM	Follow
5	1/2/23 9:00 AM	Like

Figure 4.28 – User activity table

The company wants to retrieve the activities that were performed on the current date and at the current time. It would use the CURRENT_TIMESTAMP(), NOW(), or SYSDATE() function in a SQL query to retrieve the current date and time and then use that date and time in a WHERE clause to filter the user_activity table to only the activities performed on the current date and at the current time.

Here's an example of a SQL query that would accomplish this using the CURRENT_TIMESTAMP () function:

```
SELECT user_id, activity_type FROM user_activity WHERE date_time =
CURRENT_TIMESTAMP();
```

This query would retrieve the user ID and activity type of all the activities performed on the current date and at the current time. The CURRENT_TIMESTAMP () function is used in the WHERE clause to filter the user_activity table to only the rows where the date_time column is equal to the current date and time.

Alternatively, the same query can be written using the NOW () function:

```
SELECT user_id, activity_type FROM user_activity WHERE date_time =
NOW();
```

Alternatively, we use this:

```
SELECT user_id, activity_type FROM user_activity WHERE date_time =
SYSDATE();
```

In this scenario, the CURRENT_TIMESTAMP (), NOW (), and SYSDATE () functions are used to retrieve the current date and time and filter the user_activity table to only the rows that have the same date and time. This can be used for real-time monitoring of user activity on the platform and for troubleshooting and investigating specific events.

Case scenario 2

In the airline industry, the CURRENT_TIMESTAMP (), NOW (), and SYSDATE () functions can be used in a variety of ways to retrieve and update information in a database. Here's a scenario.

An airline wants to track the current status of flights. It has a database with a table called flights that contains information about each flight, including the flight number, departure and arrival times, and current status.

flight_number	departure_time	arrival_time	departure_airport	arrival_airport	status
AA100	9:00 AM	11:30 AM	JFK	LAX	On-Time
UA200	10:30 AM	1:00 PM	ORD	SFO	Delayed
DL300	12:00 PM	2:30 PM	ATL	DFW	Cancelled
SW400	3:00 PM	4:30 PM	BWI	MCO	On-Time
FR500	5:00 PM	7:00 PM	LHR	CDG	On-Time

Figure 4.29 – Flights table

The airline wants to update the status of a flight to Delayed if the departure time is delayed by more than 15 minutes. It would use the CURRENT_TIMESTAMP(), NOW(), or SYSDATE() function in a SQL query to retrieve the current date and time and then use that value in a CASE statement to update the status of the flight based on a comparison of the departure time and the current time:

```
UPDATE flights
SET status =
CASE
WHEN departure_time < DATE_SUB(CURRENT_TIMESTAMP(), INTERVAL 15
MINUTE) THEN 'delayed'
ELSE status
END
```

This query would update the status of flights to 'delayed' if the departure time is less than 15 minutes from the current time. The CURRENT_TIMESTAMP() function is used to retrieve the current date and time, and the DATE_SUB() function is used to calculate the time 15 minutes ago. The query checks the departure time of the flight with the calculated time, and if it is greater than 15 minutes ago, then the status is updated to Delayed. Otherwise, it stays as it is, which in this case is On-Time. This can help the airline industry to update flight statuses and notify affected customers.

CAST()

The CAST() function in SQL is used to convert one data type into another. It is one of the most used functions in the real world by data scientists and analysts to manipulate data types into the right data type format.

Here's the syntax:

```
CAST (expression AS data_type)
```

Here, expression is the value or column that you want to convert and data_type is the desired data type of the output.

Here's an example.

If you had a column named price with a data type of INT and you wanted to convert it into a FLOAT data type, you would use the following SQL statement:

```
SELECT CAST(price AS FLOAT) FROM products;
```

This would return the values in the price column as the FLOAT data type.

The CAST function can also be used to convert a string into a date or a time data type. For example, to convert a string in the format yyyy-mm-dd into a DATE data type, you would use the following SQL statement:

```
SELECT CAST('2022-01-01' AS DATE);
```

This would return the string as the DATE data type 2022-01-01.

It is also possible to use the CAST function with mathematical operations, for example:

```
SELECT CAST(5.5 AS INT);
```

This would return the value 5.

It is important to note that the CAST function may truncate or round off the results when converting data types if the values do not fit into the desired data type.

Case scenario 1

One example of a case study for using the CAST() function in SQL could be in a retail company that has a database containing information about products and sales.

product_id	product_name	category	price	quantity
P001	T-Shirt	Clothing	$20	100
P002	Jeans	Clothing	$40	50
P003	Sneakers	Shoes	$60	75
P004	Hat	Accessories	$15	200
P005	Backpack	Bags	$30	150

Figure 4.30 – Products table

The company might have a column in the products table called price that is currently stored as an INT data type, but the company wants to perform calculations on the prices that require the use of decimal places, such as calculating the average price of all products.

In this case, the company would use the CAST() function to convert the price column from the INT into the FLOAT data type so that decimal places can be used in calculations.

The SQL statement would be the following:

```
SELECT CAST(price AS FLOAT) FROM products;
```

This would return the values in the price column as the FLOAT data type, allowing the company to perform calculations that require decimal places.

Case scenario 2

A company wants to track the time of purchase, but the time is stored as a string in the format yyyy-mm-dd hh:mm:ss in the sales table. The company wants to extract the time of purchase in a specific time format to perform analysis based on the time of purchase.

product_id	product_name	category	price	quantity	time_of_purchase
P001	T-Shirt	Clothing	$20	100	1/1/2023 10:00AM
P002	Jeans	Clothing	$40	50	1/1/23 11:00 AM
P003	Sneakers	Shoes	$60	75	1/1/23 12:00 PM
P004	Hat	Accessories	$15	200	1/1/23 2:00 PM
P005	Backpack	Bags	$30	150	1/2/23 9:00 AM

Figure 4.31 – Sales table

In this case, the company would use the CAST() function to convert the string into a time data type.

The SQL statement would be the following:

```
SELECT CAST(time_of_purchase AS TIME) FROM sales;
```

This would return the time of purchase as a time data type, allowing the company to perform analysis and extract specific information about the time of purchase.

In summary, the CAST() function allows you to convert a data type into another one. This is useful when performing calculations or analyses that require a specific data type, and it can be applied in many scenarios in any company that uses a SQL database.

Benefits of the CAST function

Let's look at the benefits of using a CAST() function:

- Changing the data type of a column to facilitate data analysis or reporting
- Converting a string value into a numeric value for mathematical operations
- Converting a date column into a string column for better readability or string operations
- Converting a timestamp into a date type to group by date
- Converting a numeric value into a character string for concatenation or string manipulation
- Converting a NULL value into a specific data type to avoid errors while performing operations

Summary

This brings us to the end of this chapter, and by now, you should have learned about the following:

- The different ways to store date and time information in a SQL database, specifically the DATETIME, DATE, and TIME data types
- The syntax and format for each of these data types

- Various built-in functions provided by SQL for manipulating and extracting information from DateTime data, such as NOW(), DATE(), and DATEDIFF()

- How to perform calculations and comparisons with DateTime data using SQL functions

- The importance of good data validation and integrity in order to ensure that the data stored in DateTime columns is accurate, consistent, and in the correct format

In the next chapter, we will learn about Null values in a dataset and will look at ways to avoid and handle such values in a dataset. Handling null values in SQL is an important step before introducing data wrangling, as it ensures that your data is accurate, consistent, and ready for data wrangling and analysis. This will help you avoid any errors and get better insights from your data and avoid an impact on the final outcome of the analysis.

Handling NULL Values

In this chapter, we will delve into the crucial aspects of dealing with missing data and NULL values during the data analysis process. We will explore the profound impact that these gaps in data can have on the accuracy and validity of our analyses if not appropriately addressed. By emphasizing the significance of data validation and cleaning prior to analysis, we will highlight the importance of ensuring accurate, complete, and relevant data in drawing meaningful conclusions and facilitating informed decision-making.

We will begin by unraveling the distinction between NULL values and 0, shedding light on their different meanings within the context of data analysis. Recognizing this distinction serves as the foundation for comprehending the different ways to handle NULL values effectively. Throughout this chapter, we will outline various methods for managing missing data in SQL, empowering you to make informed decisions when dealing with these data gaps.

By undertaking data validation, we will establish a systematic process for identifying and validating missing data or NULL values. We will emphasize the significance of this process as a precursor to data cleaning techniques, which aim to remove or replace the gaps in the dataset. The repercussions of drawing incorrect conclusions and hindering effective decision-making due to inaccurate, incomplete, or irrelevant data are extensively explored. Handling missing data and NULL values not only ensures data accuracy but also enhances the overall performance of the data analysis process.

With this comprehensive introduction, let's learn about the tools required to navigate the challenges posed by missing data and NULL values. By emphasizing the crucial role of data validation and cleaning, we will set the stage for effective data analysis that yields reliable insights and facilitates informed decision-making.

In this chapter, we will cover the following main topics:

- The impact of missing data or NULL values on data analysis
- Understanding the importance of data validation and cleaning before analyzing data
- The difference between NULL values and 0 values
- Different ways to handle NULL values

The impact of missing data and NULL values on data analysis

Missing data or NULL values can have a significant impact on data analysis. When data is missing or NULL, it can lead to incorrect conclusions and poor decision-making. This is because missing data and NULL values can introduce bias and uncertainty into the analysis, and lead to a loss of information and reduced power to detect meaningful patterns in the data.

One of the main impacts of missing data and NULL values is that if they are not handled properly, they can skew the results of the analysis by introducing systematic errors or distorting the distribution of the data. For example, imagine you are conducting a survey to study the relationship between income levels and home ownership. You collect data on variables such as income, age, education level, and homeownership status. However, for various reasons, some respondents fail to provide their homeownership status, resulting in missing values for that particular variable.

Now, let's say you proceed with the analysis without addressing these missing values. Any conclusions or insights drawn from the dataset will only reflect the observations of respondents who provided their homeownership status. The analysis would inherently be biased toward this subset of respondents, potentially leading to misleading findings.

For instance, if you find a strong positive correlation between income and homeownership in the available data, it might not accurately represent the entire population. It is possible that the missing data disproportionately includes lower-income individuals who might have a different homeownership pattern. Consequently, the conclusion drawn from the analysis would be skewed and not reflective of the true relationship between income and homeownership in the broader population.

Therefore, it becomes crucial to handle missing values appropriately to prevent biased or inaccurate conclusions. By addressing the missing data through data validation and cleaning techniques, you can ensure that the analysis is based on a representative data sample, providing more reliable insights into the relationship between income and homeownership.

Similarly, let's try and understand the impact of NULL values on a dataset using an example. Let's say you work for an e-commerce company and are analyzing customer purchase data. The database contains a table with customer information, including their purchase amounts. However, due to occasional technical issues during the data collection process, some of the purchase amounts are recorded as NULL values.

If you were to calculate the average purchase amount for all customers without handling the NULL values, the result would be affected. The SQL query might return a NULL value for the average instead of a meaningful value. This occurs because by default, NULL values are not taken into account in mathematical calculations.

If you were to use this average purchase amount as a **key performance indicator** (**KPI**) or use it to make business decisions, it could lead to incorrect insights or flawed decision-making. The missing values could introduce bias and undermine the accuracy of your analysis.

To overcome this issue, it is essential to handle NULL values appropriately during SQL analysis. There are different approaches you can take, such as excluding NULL values from calculations or replacing them with a different value (e.g., 0) before performing calculations.

By properly handling the NULL values, you ensure that your analysis is based on complete and accurate data, enabling you to make more informed business decisions. It also helps prevent the potential pitfalls and inaccuracies that can arise from incomplete or missing information in the dataset.

Understanding the importance of data validation and cleaning before analyzing data

Data validation and cleaning are particularly important when working with missing data and NULL values in SQL. Missing data and NULL values can have a significant impact on the accuracy and validity of analysis if not handled properly.

In SQL, data validation involves checking for missing data and NULL values in the dataset and determining whether they are legitimate missing values or whether they indicate a data entry error. For example, a NULL value in the `customer_email` column might indicate that the customer did not provide an email address, or it might indicate that the data entry person made an error and forgot to enter the email address. By identifying and validating missing data and NULL values, you can ensure that the dataset is accurate and complete.

Data cleaning with respect to missing data and NULL values in SQL involves removing or replacing missing data and NULL values. This can be done in a number of ways, depending on the nature of the missing data and NULL values and the goals of the analysis:

- **Listwise deletion**: This is the most common and simplest approach and involves removing any rows or columns that contain missing data or NULL values. This can help to ensure that the dataset is accurate, but it can also lead to a loss of data and bias in the analysis.

- **Imputation**: This method involves replacing missing data or NULL values with a substitute value, such as the mean or median of the column. This can help to preserve the data and prevent bias in the analysis, but it can also introduce errors if the imputed value is not representative of the true value.

- **Single imputation**: This is a specific type of imputation where missing data is filled with a single value. This could be a constant value or a statistical value such as the mean, median, or mode.

- **Multiple imputation**: This is a more advanced approach that involves using machine learning techniques to estimate the missing values. This can be more accurate and efficient than other methods, but it also requires more computational resources and domain knowledge.

- **Using NULL**: This method involves leaving missing data as NULL values in the dataset and using SQL's built-in handling of NULL values to filter or aggregate the data as needed.

- **Using NULL with outer join**: This method involves using an outer join to combine tables and using the NULL values to identify and exclude missing data.

The choice of method will depend on the nature of the data, the goals of the analysis, and the resources available. It's important to evaluate the methods and choose the one that's the most appropriate for the research question or business problem at hand.

Identifying NULL/missing values

Let's consider an example table called `customers` to demonstrate how to identify and count NULL or missing values using the `COUNT()` function in SQL.

Here's an example table called `customers`:

customer_id	email
1	john@example.com
2	NULL
3	sarah@example.com
4	NULL
5	mike@example.com
6	NULL

Figure 5.1 – customers table

To identify and count the occurrences of NULL or missing values in the `email` column of the `customers` table, you can use the following SQL query:

```
SELECT COUNT(*) AS NULL_count FROM customers WHERE email IS NULL;
```

The query will return a single row with a column called `NULL_count`, which indicates the number of NULL or missing values found in the `email` column.

In this example, the result of the query is as follows:

NULL_count
3

Figure 5.2 – Query output

The result shows that there are three NULL or missing values in the `email` column of the `customers` table.

By using the COUNT() function in conjunction with the IS NULL condition and the WHERE clause, you can accurately identify and count the occurrences of NULL and missing values in a specific column. This allows you to gain insights into the amount of missing data in your dataset, which can be helpful for data cleaning, analysis, and decision-making.

NULL values versus zero values

In this section, we will learn how NULL values are different from zero values.

In SQL, a NULL value represents the absence of a value, or unknown data. It is different from an empty value and a zero value, which have specific meanings.

An empty value, also known as a blank or empty string, is a value that has no characters or whitespace. In SQL, an empty string is represented by two single quotes with nothing in between, as follows:

```
select * from vendor
where vendor_email = ''
```

In this example, single quotes are used without any gap in between.

A zero value, on the other hand, is a numerical value that represents the absence of a value or a known quantity. In SQL, a zero value is represented by the number 0.

It's important to understand that NULL values are not the same as empty or zero values. A NULL value indicates that the data is missing or unknown, whereas an empty and zero values have specific meanings.

When working with NULL values in SQL, it's important to use the IS NULL and IS NOT NULL operators to filter and select data with NULL values because, due to the specific format of NULL values, it is not possible to use traditional comparison (=, <, >, and <>) operators in the queries. These operators can be used in the WHERE clause of a SELECT, UPDATE, or DELETE statement to filter the data based on whether a column contains NULL values or not.

For example, the following query will return all rows from the customers table where the value in the phone column is NULL:

```
SELECT * FROM customers WHERE phone IS NULL;
```

It's also important to use the COALESCE and IFNULL functions to replace NULL values with default values. These functions can be used in the SELECT clause of a query to return a default value when a column or expression is NULL.

For example, the following query will return the phone number of a customer, or a default value of N/A if the phone number is NULL:

```
SELECT name, COALESCE(phone, 'N/A') AS phone FROM customers;
```

It's important to keep in mind that NULL values can have a significant impact on the results of a query and should be handled properly to ensure accurate and meaningful results.

Using the IS NULL and IS NOT NULL operators to filter and select data with NULL values

The IS NULL and IS NOT NULL operators in SQL are used to filter and select data with NULL values. These operators are placed in the WHERE clause of SELECT, UPDATE, and DELETE statements to filter the data based on whether a column has NULL or non-NULL values. Specifically, the IS NULL operator can be used to filter out rows where a particular column has a NULL value, while the IS NOT NULL operator can be used to filter out rows where a particular column doesn't have a NULL value.

IS NULL() and IS NOT NULL() – scenario

A real-world scenario where the IS NULL and IS NOT NULL operators are useful is in the transportation industry. Say a company wants to track the maintenance status of their fleet of vehicles but not all vehicles have complete maintenance records.

They have a table named vehicles with the following data:

make	model	last_service	next_service
Audi	A4	1/1/2022	1/1/2023
Mercedes	C-Class	1/2/2022	1/2/2023
BMW	3-Series	NULL	1/3/2023
Ford	Mustang	1/4/2022	NULL
Honda	Civic	1/5/2022	1/5/2023

Figure 5.3 – vehicles table

In this table, not all vehicles have complete maintenance records with the next service date.

The company can use the IS NULL operator to filter vehicles that are overdue for a service:

```
SELECT
    make,
    model,
    last_service,
    next_service
FROM vehicles
WHERE next_service IS NULL;
```

This query will return the following result; as you can see, next_service is NULL:

make	model	last_service	next_service
Ford	Mustang	1/4/2022	NULL

Figure 5.4 – Query result

Now, if the company also wants to know the details of all the cars that have been serviced, they can use IS NOT NULL as follows:

```
SELECT
    make,
    model,
    last_service,
    next_service
FROM vehicles
WHERE next_service IS NOT NULL;
```

This query will return the following result. As you can see, next_service is not NULL and all the cars here have been serviced:

make	model	last_service	next_service
Audi	A4	1/1/2022	1/1/2023
Mercedes	C-Class	1/2/2022	1/2/2023
BMW	3-Series	NULL	1/3/2023
Honda	Civic	1/5/2022	1/5/2023

Figure 5.5 – Query result

Here's an interesting industry example of how IS NOT NULL and IS NULL can be used in SQL in the retail industry.

A retail company maintains a database containing customer information and purchase histories. Their objective is to identify customers who have never made a purchase, enabling them to target them with promotional offers. There are multiple scenarios where this can occur. For example, consider the following:

- A company might possess this data based on survey databases that capture information from conducted surveys

- A customer could have visited the website without making a purchase, even though they are a registered member

To do this, they use the `IS NULL` statement to find all customers whose `first_purchase_date` column is NULL. This returns a list of all customers who have never made a purchase.

On the other hand, the company also wants to find out which customers have made a purchase but haven't returned any items. They use the `IS NOT` statement to find all customers whose `returns` column is not equal to 0. This returns a list of all customers who have made a purchase but haven't returned any items.

The company can use this information to send these customers special offers or incentives to encourage them to make a purchase.

These operators can be used together with other operators or functions to filter and select the required data. These operators are useful in situations where a table may not have complete data and it's important to filter or select only the rows that contain the required data.

Why can't we use the returns > 0 condition in this case?

In our example, we are trying to identify customers who have made a purchase but haven't returned any items. While it might seem logical to use the `returns > 0` condition to achieve this, it can lead to incorrect results or exclude certain cases.

Using `returns > 0` would indeed filter out customers who have returned at least one item, but it could also exclude customers who have not made any returns yet. This is because the `returns > 0` condition would only include customers with a positive value in the `returns` column, omitting those with a NULL value or zero returns.

If there are customers who have made purchases but haven't returned any items yet, they would have a NULL or zero value in the `returns` column. Using `returns > 0` as the condition would mistakenly exclude those customers from the results.

To ensure we include customers who have made a purchase but haven't returned any items, it is more appropriate to use `returns IS NOT 0` or `returns IS NULL`, depending on the data representation. This allows us to capture customers with NULL or zero values in the `returns` column, providing a comprehensive view of the target customer segment.

By using the correct condition, we can accurately identify customers who have made purchases without any returns, enabling the retail company to tailor their strategies and incentives accordingly.

Using the COALESCE and IFNULL functions to replace NULL values with a default value

When working with databases, the `COALESCE` and `IFNULL` functions serve as valuable tools for replacing NULL values with predefined defaults. These functions help ensure data consistency and improve query results by providing a fallback option when encountering NULL values in the database.

IFNULL()

The SQL IFNULL function is used to return a specified value if the expression is NULL.

The basic syntax for using IFNULL is as follows:

```
IFNULL(expression, replacement_value)
```

Here, expression is the value that you want to check for NULL values, and replacement_value is the value that you want to return if expression is NULL.

For example, the following query will return N/A if the value in the name column is NULL:

```
SELECT IFNULL(name, 'N/A') FROM table_name;
```

This will result in missing values returning N/A.

This is an alternative:

```
SELECT IFNULL(name, 0) FROM table_name;
```

This will result in missing values returning 0.

> **Note**
> This is an example of MySQL syntax; other DBMSes may have different syntax.

Scenario

A real-world scenario where the IFNULL function is useful is in a retail database where you want to track customer orders. In this database, you have a table called orders that contains information about each order, including customer_ID and total_sales.

customer_ID	total_sales
1	$5,000
2	$2,000
NULL	$3,000
3	$1,500
NULL	$500

Figure 5.6 – orders table

However, not all customers provide their information when placing an order (for example, a customer may make a purchase in store without providing their contact information). In this case, the customer_ID column in the orders table will contain NULL values for these orders.

You want to create a report that shows the total sales by customer, but you don't want to exclude the orders where `customer_ID` is missing.

You can use the `IFNULL` function to replace the NULL values with a default value, such as `anonymous`, so that these orders will be included in the report and grouped under `anonymous`

The query would look something like this:

```
SELECT
    IFNULL(customer_id, 'anonymous') as customer,
    SUM(total) as total_sales
FROM orders
GROUP BY customer;
```

This query returns a report that shows the total sales for each customer, including orders where the `customer_id` is missing.

customer_id	total_sales
1	$5,000
2	$2,000
3	$1,500
Anonymous	$3,500

Figure 5.7 – Query result

This allows you to get a complete picture of the sales even when some data is missing and allows you to make better business decisions.

An interesting industry example of how the `IFNULL` function can be used in SQL is in the banking industry, where banks have a large amount of customer data that needs to be analyzed to make informed business decisions.

Imagine that a bank wants to analyze the spending patterns of their customers and, as part of this analysis, they want to group customers by their age. However, some customers do not provide their age when they open an account, so the `age` column in the customer table contains NULL values.

The bank can use the `IFNULL` function to assign a default age value to these customers, such as `unknown`, so that they can be included in the analysis.

The following query can be used to group customers by age and calculate the total spending for each age group:

```
SELECT
    IFNULL(age, 'unknown') as age,
    SUM(spending) as total_spending
```

```
FROM customer
GROUP BY age;
```

This query returns a report that shows the total spending for each age group, including customers whose ages are unknown.

age	total_spending
18-24	$2,000
25-34	$4,000
35-44	$1,500
45-54	$2,000
55+	$500
Unknown	$800

Figure 5.8 – Query result

This way, the bank can get a better understanding of the spending patterns of their customers, even when some data is missing. They can also track the spending of an unknown age group and compare it with a known age group, which will help them to target their marketing strategy.

Several key metrics and insights can be extracted from the information in the example I provided:

- **Age distribution of customers**: The bank can see the distribution of customers by age, including the percentage of customers whose age is unknown. This can help the bank understand the demographic makeup of its customer base.

- **Total spending by age group**: The bank can see the total spending for each age group, which can give it an idea of the spending patterns of customers at different stages of life.

- **Average spending by age group**: The bank can calculate the average spending per customer for each age group by dividing the total spending by the number of customers in that group. This can give the bank an idea of how much each customer in a specific age group spends on average.

- **Unknown age group spending**: The bank can track the spending of the unknown age group and compare it with the known age group, which can help it target its marketing strategy more effectively.

- **Identifying the age group with the highest spending**: The bank can identify which age group has the highest spending, which can help it focus its marketing efforts on that group to attract more customers of the same age group.

- **Identifying the age group with the lowest spending**: The bank can also identify which age group has the lowest spending, which can help it to focus on that group to find out why they are not spending as much as other groups and try to improve their spending.

Next, we look at the COALESCE function.

COALESCE()

The COALESCE() function and the IFNULL() function work in a similar way. The COALESCE() function is used to return the first non-NULL value in a list of expressions. The syntax for the function is as follows:

```
COALESCE(expression1, expression2, ... expression_n);
```

The function returns the first non-NULL expression in the list. If all expressions are NULL, the function returns NULL. It can be used to return a default value when a column or expression is NULL by passing the default value as the second argument.

For example, the following query will return the value 0 if the price column is NULL:

```
SELECT name, COALESCE(price, 0) AS price FROM orders;
```

This will return 0 for missing values because we are using 0 where the price is NULL.

Scenario

An interesting use case for the COALESCE() function in a real-world scenario is in a **customer relationship management (CRM)** system. The system needs to display a customer's preferred contact method, but the customer may not have provided one.

For example, consider a table named customers with the following data:

id	name	email	phone
1	Dave Smith	dave.smith@gmail.com	8655-985-5555
2	Joseph Doe	joseph.doe@yahoo.com	8756-542-5256
3	Mike S	mike.s@gmail.com	NULL
4	Sam Berlin	sam.b@gmail.com	523-455-5558
5	Steven	NULL	585-555-5559

Figure 5.9 – customers table

In this table, the system wants to display the customer's preferred contact method, email, or phone. But some of the customers did not provide a contact method:

```
SELECT name, COALESCE(email, phone) AS contact_method FROM customers;
```

This query will return the following result:

name	contact_method
Dave Smith	`dave.smith@gmail.com`
Joseph Doe	`joseph.doe@yahoo.com`
Mike S	`mike.s@gmail.com`
Sam Berlin	`sam.b@gmail.com`
Steven	585-555-5559

Figure 5.10 – Query result

In this example, the COALESCE () function is used to return the email or phone of a customer, if the customer provided one. If the customer did not provide a contact method, the function returns NULL. This allows the system to display the customer's preferred contact method, even if the customer did not provide one.

This is useful in scenarios where a company wants to reach out to its customers, but not all customers have provided contact information. The COALESCE function can be used to check multiple columns and return the first non-NULL value, ensuring that the company can still contact the customer.

An interesting industry example of how the COALESCE () function can be used in SQL for social media analytics is in a scenario where a company wants to track the engagement of their social media posts, but not all posts have the same type of engagement data.

For example, consider a table named posts with the following data:

Id	post	likes	shares
1	"Hello World"	150	20
2	"How to make a cake"	100	15
3	"Technology updates"	NULL	10
4	"COVID-19 prevention"	60	NULL
5	"Mental Health"	80	20

Figure 5.11 – posts table

This table represents a sample of five posts, where id is the unique identifier of each post, post is the post's content, likes is the number of likes on the post, and shares is the number of shares of the post. As you can see, there are NULL values in the likes and shares columns for posts with the IDs 3 and 4.

The company can use the COALESCE () function to display the engagement data for each post, or a default value of 0 if the data is not available:

```
SELECT
    post,
    COALESCE(likes, 0) AS likes,
    COALESCE(shares, 0) AS shares
FROM posts;
```

This query will return the following result:

Post	Likes	shares
"Hello World"	150	20
"How to make a cake"	100	15
"Technology updates"	0	10
"COVID-19 prevention"	60	0
"Mental Health"	80	20

Figure 5.12 – Query result

With this query, the company can easily track the engagement data for each post, even if some posts do not have data for likes or shares.

This is useful in scenarios where a company wants to track engagement data across multiple social media platforms and not all platforms provide the same information. The COALESCE function can be used to check multiple columns and return a default value, ensuring that the company can still analyze the engagement data.

There are several key metrics and insights that can be extracted from the information in the example I provided:

- **Engagement rate**: The engagement rate can be calculated by dividing the number of likes, shares, and comments by the number of followers or reach, and multiplying the result by 100. This metric can be used to measure how engaged a post's audience is.

- **Virality**: The virality of a post can be measured by the number of shares. A post with a high number of shares is considered viral.

> **Note**
>
> In Oracle, the `ISNULL` function is used, and in other databases, such as MS SQL and DB2, the NVL function is used. These functions are used to handle missing or NULL values and to ensure that the data is complete and accurate for analysis and reporting. In PostgreSQL, the equivalent function for handling NULL values is called `COALESCE`. The `COALESCE` function allows you to replace NULL values with an alternative non-NULL value. They allow robust and effective data analysis even when some data is missing.

In summary, `IFNULL`, `ISNULL`, NVL, and `COALESCE` are functions in SQL that are used to handle NULL values and return a specified value when the input is NULL. They are used to make sure the data is complete and accurate for analysis and reporting.

IS NULL versus = NULL

The difference between `IS NULL` and `= NULL` is an important distinction in SQL and can lead to different outcomes when dealing with NULL values. Let's dive into the differences between them.

Let's consider an example table dataset called `employees` to demonstrate the difference between `IS NULL` and `= NULL`.

employee_id	employee_name	salary
1	John	5000
2	Sarah	NULL
3	Mike	3000

Figure 5.13 – employees table

Now, let's examine the differences in behavior when using `IS NULL` and `= NULL` to check for NULL values in the `salary` column:

- Using `IS NULL`:

  ```
  SELECT * FROM employees WHERE salary IS NULL;
  ```

  This query will return the row where `employee_id` is 2, where the `salary` column is NULL.

- Using `= NULL`:

  ```
  SELECT * FROM employees WHERE salary = NULL;
  ```

  This query will not return any rows, even though we have a NULL value in the `salary` column. This is because comparing a value to NULL using the equality operator (=) always results in NULL, not `true` or `false`.

To illustrate this behavior further, let's add another row to the `employees` table:

employee_id	employee_name	salary
1	John	5000
2	Sarah	NULL
3	Mike	3000
4	Jane	NULL

Figure 5.14 – employees table with an extra row

Now, let's run the same queries again:

- Using `IS NULL`:

  ```
  SELECT * FROM employees WHERE salary IS NULL;
  ```

  The query will return the rows for `employee_id` 2 and 4, where the `salary` column is NULL.

- Using `= NULL`:

  ```
  SELECT * FROM employees WHERE salary = NULL;
  ```

  The query will not return any rows, even though we have two rows with NULL values in the `salary` column.

From these examples, it becomes evident that `IS NULL` should be used to check for NULL values in SQL, while `= NULL` does not yield the expected result and should be avoided.

Summary

This chapter covered the importance and different ways of handling NULL values and missing values in SQL. It's essential to identify and handle missing data or NULL values in a way that preserves the integrity of the analysis. The different methods discussed in this chapter can be chosen based on the nature of the data and the goals of the analysis.

In conclusion, important topics that were covered in this chapter are as follows:

- The impact of missing data or NULL values on data analysis
- Different ways to handle missing data or NULL values in SQL
- The importance of data validation and cleaning

Now that we have a good understanding of how to handle missing data and NULL values in SQL, it's time to move on to the next chapter and learn about another important topic: pivoting data using SQL.

6

Pivoting Data Using SQL

Pivoting data in SQL is useful for data wrangling because it allows you to reshape the structure of your data in a way that is more useful for analysis and visualization. Some reasons why you might want to `pivot` data include the following:

- **Presenting data in a more intuitive format**: Pivoting data can make it easier to understand and interpret by transforming it into a more familiar layout, such as a table with rows and columns

- **Facilitating data analysis**: Pivoting data can make it easier to perform certain types of analysis, such as comparing the values of different groups or calculating aggregate values

- **Enhancing data visualization**: Pivoting data can make it easier to create visualizations that are more meaningful and informative, such as bar charts, line graphs, and heat maps

- **Creating a consistent data structure**: Pivoting data can be used to transform data into a consistent format, making it easier to join with other data sources, perform data quality checks, and automate reports

- **Reducing the size of the dataset**: Pivoting data can also be used to reduce the size of the dataset, as it allows you to aggregate data, removing unnecessary data and simplifying the data structure.

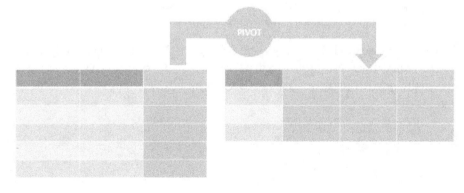

Figure 6.1 – Pivot diagram

Overall, pivoting data in SQL allows for more efficient data manipulation and analysis, making it a useful tool for data wrangling. Let's now see how can we achieve this using various in-built functions of SQL Server.

In this chapter, we will cover the following main topics:

- SQL Transpose – rows to columns
- SQL Cross Tab – columns to rows
- Analytical workflow – from SQL to business intelligence – transforming data into actionable insights

SQL Transpose – rows to columns

In SQL, there are several ways to transpose data from rows to columns, also known as "horizontal to vertical" or "flipping" the data. One common way is to use the `PIVOT` function.

The `PIVOT` function allows you to rotate rows into columns, effectively transposing the data.

The basic syntax for using the `PIVOT` function is as follows:

```
SELECT ..., [column_to_become_new_column_1], [column_to_become_new_
column_2], ...
FROM ...
PIVOT (aggregate_function(column_to_aggregate)
     FOR column_to_become_new_columns
     IN ([list_of_values_to_become_new_columns]))
```

In this syntax, `...` represents any additional columns that you want to include in the output, `aggregate_function` is the function used to aggregate the values in the column that you want to transpose, `column_to_aggregate` is the name of the column that you want to transpose, `column_to_become_new_columns` is the name of the column whose values will become the new columns in the output, and `[list_of_values_to_become_new_columns]` is a list of the values that you want to use as column names in the output.

For example, if you have a table called `sales` with the "`product`", "`month`", and "`sales_amount`" columns and you want to transpose the data so that the values in the "`month`" column becomes the new column and the "`sales_amount`" column is aggregated by the sum, you will use the query provided after the table:

product	month	sales_amount
A	January	100
A	February	150
A	March	200
A	April	180
B	January	80
B	February	120
B	March	90
B	April	110
C	January	50
C	February	60
C	March	70
C	April	80
D	January	70
D	February	90
D	March	110
D	April	100

Figure 6.2 – sales table

```
SELECT product, [January], [February], [March], [April]
FROM sales
PIVOT (SUM(sales_amount)
      FOR month
      IN ([January], [February], [March], [April]))
```

This would produce a new table with a column for each month and the sum of the sales amounts for that month.

You can also use a subquery in the IN clause if the values to be used as columns are not known in advance:

```
SELECT product, [January], [February], [March], [April]
FROM sales
PIVOT (SUM(sales_amount)
      FOR month
      IN (SELECT DISTINCT month FROM sales))
```

product	January	February	March	April
A	100	150	200	180
B	80	120	90	110
C	50	60	70	80
D	70	90	110	100

Figure 6.3 – Query result

> **Note**
>
> PIVOT function support may vary depending on the **database management system (DBMS)** you are using.
>
> Additionally, the order of the months in the preceding subquery (IN (SELECT DISTINCT month FROM sales) depends on the sequence in which they were created in the sales table, that is, January, February, March, and so on.

Let's move on to some real-world use case scenarios.

Use case scenario

Suppose you have a table called patient_data that stores information about patients in a hospital, and it has the following columns:

- patient_id: A unique identifier for the patient

- test_type: The type of medical test performed (e.g., blood test, X-ray, etc.)

- test_result: The result of the test (e.g., normal, abnormal, etc.)

- test_date: The date the test was performed

patient_id	test_type	test_result	test_date
1	Blood Test	Normal	1/1/2021
1	X-ray	Abnormal	1/2/2021
1	CT scan	Normal	1/3/2021
2	Blood Test	Abnormal	1/1/2021
2	X-ray	Normal	1/2/2021
2	CT scan	Abnormal	1/3/2021

Figure 6.4 – patient_data table

You want to create a new table that shows the test results for each patient, with the test type as the columns and the test result as the values.

Here's an example of how you can use the PIVOT function to achieve this:

```
SELECT patient_id, [Blood_Test], [X_ray], [CT_scan]
FROM patient_data
PIVOT (MAX(test_result)
      FOR test_type
      IN ([Blood Test], [X-ray], [CT scan]))
```

This query will give you a new table with columns for each test type and the corresponding test result for each patient.

patient_id	Blood_Test	X_ray	CT_scan
1	Normal	Abnormal	Normal
2	Abnormal	Normal	Abnormal

Figure 6.5 – Query result

This way, you can easily and quickly pivot your data to make it more readable and understandable.

You can also use aggregate functions such as COUNT, SUM, or AVG and a GROUP BY clause with the PIVOT function to get the desired result based on your requirements.

It's worth noting that this query will only return rows for the test types specified in the pivot. If you have test types that are not known in advance or you want to include all test types, you can use a subquery in the IN clause:

```
SELECT patient_id, [Blood Test], [X-ray], [CT scan]
FROM patient_data
PIVOT (MAX(test_result)
      FOR test_type
      IN (SELECT DISTINCT test_type FROM patient_data))
```

This query will give you a new table with columns for each test type and the corresponding test result for each patient, regardless of the test types that are present in the data.

Let's see how the PIVOT function is used in a real-world industry scenario to analyze data and make informed decisions. Suppose you have a table called delivery_data that stores information about deliveries made by a transportation company, and it has the following columns:

- delivery_id: A unique identifier for the delivery
- delivery_date: The date the delivery was made

- `route`: The route taken for the delivery
- `delivery_status`: The status of the delivery (e.g., Delivered, Delayed, or Cancelled)
- `vehicle_type`: The type of vehicle used for the delivery (e.g., truck, van, or bike)

delivery_id	delivery_date	route	delivery_status	vehicle_type
1	1/1/2021	Route A	Delivered	truck
2	1/1/2021	Route A	Delayed	van
3	1/2/2021	Route B	Delivered	bike
4	1/2/2021	Route B	Delivered	truck
5	1/3/2021	Route A	Cancelled	van

Figure 6.6 – delivery_data table

You want to create a new table that shows the delivery status for each route, with the vehicle type as the columns and the delivery status as the values.

Here's an example of how you can use the PIVOT function to achieve this:

```
SELECT route, [truck], [van], [bike]
FROM delivery_data
PIVOT (COUNT(delivery_status)
      FOR vehicle_type
      IN ([truck], [van], [bike]))
GROUP BY route
```

This query will give you a new table with columns for each vehicle type and the number of deliveries made with that vehicle type for each route.

route	truck	van	bike
A	1	2	0
B	1	0	1

Figure 6.7 – Query result

You can also use aggregate functions such as sum, avg, max, and min with the pivot function to get the desired result based on your requirements.

It's worth noting that this query will only return rows for the vehicle types specified in `pivot`. If you have vehicle types that are not known in advance or you want to include all vehicle types, you can use a subquery in the `IN` clause. Try to write this query on your own before looking at the solution here:

```
SELECT route, [truck], [van], [bike]
FROM delivery_data
PIVOT (COUNT(delivery_status)
       FOR vehicle_type
       IN (SELECT DISTINCT vehicle_type FROM delivery_data))
GROUP BY route
```

This query will give you a new table with columns for each vehicle type and the number of deliveries made with that vehicle type for each route, regardless of the vehicle types that are present in the data.

This way, you can easily and quickly `pivot` your data to make it more readable and understandable, giving the company insights into the delivery status and vehicle type used for each route.

SQL Cross Tab – columns to rows

A `cross tab` or `crosstab` query is a type of query that creates a matrix-like output with the values of one column on the *x* axis and the values of another column on the *y* axis. It is also known as a `pivot` or `transpose` query. The goal of data wrangling is to transform raw data into a format that is useful for analysis, and a cross tab query is one way to do that.

Here's an example of how you might use a `crosstab` query to transform the data in the `delivery_data` table so that the rows and columns are rearranged in a way that is more meaningful for your analysis:

delivery_id	delivery_date	route	delivery_status	vehicle_type
1	1/1/2021	Route A	Delivered	truck
2	1/1/2021	Route A	Delayed	van
3	1/2/2021	Route B	Delivered	bike
4	1/2/2021	Route B	In Transit	truck
5	1/3/2021	Route A	In Transit	van

Figure 6.8 – delivery_data table

```
WITH cte AS
  SELECT route, delivery_status, COUNT(delivery_id) as count
  FROM delivery_data
  GROUP BY route, delivery_status
)
SELECT route,[Delivered],[In_Transit],[Delayed]
FROM cte
PIVOT (
  SUM(count)
  FOR delivery_status IN ('Delivered', 'In Transit',' Delayed')
)
```

This query would return a table with the following columns:

route	Delivered	In_Transit	Delayed
A	1	1	1
B	1	1	0

Figure 6.9 – Query result

As you can see, the original rows have been rearranged so that the delivery_status values are now columns, and the route values are now rows. The COUNT function is used to display the number of deliveries for each status and route combination.

Use case scenario

Imagine you work for a retail company and you want to analyze the sales data for different product categories and regions. The sales data is stored in a table called sales_data with the following columns:

- sale_id: A unique identifier for each sale
- product_category: The category of the product that was sold (e.g., electronics, clothing, or home goods)
- region: The region where the sales took place (e.g., North, South, East, or West)
- sale_amount: The amount of sales

The table is shown in *Figure 6.10*.

sale_id	product_category	region	sale_amount
1	electronics	East	$100
5	clothing	East	$60
4	electronics	North	$925
3	home goods	South	$200
2	clothing	West	$50

Figure 6.10 – sales_data table

Here's an example of how you might use the SQL GROUP BY and CASE statements to create a crosstab query:

```
WITH wrangled_data AS (
  SELECT
    product_category,
    region,
    SUM(sale_amount) as total_sale_amount
  FROM sales_data
  GROUP BY product_category, region
)
SELECT
  product_category,
  SUM(CASE WHEN region = 'North' THEN total_sale_amount ELSE 0 END) as
'NorthRegion',
  SUM(CASE WHEN region = 'South' THEN total_sale_amount ELSE 0 END) as
'SouthRegion',
  SUM(CASE WHEN region = 'East' THEN total_sale_amount ELSE 0 END) as
'EastRegion',
  SUM(CASE WHEN region = 'West' THEN total_sale_amount ELSE 0 END) as
'WestRegion'
FROM wrangled_data
GROUP BY product_category;
```

This query will give you a table with the following columns:

product_category	East	North	South	West
clothing	$ 60.00	$ -	$ -	$ 50.00
electronics	$ 100.00	$ 925.00	$ -	$ -
home goods	$ -	$ -	$ 200.00	$ -

Figure 6.11 – Query result

As you can see, the original rows have been rearranged so that the region values are now columns, and the `product_category` values are now rows. The `SUM` function is used to display the total sales amount for each `product_category` and region combination.

This query will give you a summary of the total sales amount by region and product category. You can use this data to analyze the sales data and make informed decisions about how to allocate resources and target marketing efforts.

The preceding cross tab query example can be used to derive several important insights and metrics related to sales data. Here are a few examples of the insights and metrics that you can extract from the query results:

- **Sales by product category and region**: By viewing the total sales amount for each product category and region combination, you can quickly identify which product categories and regions are performing well, and which ones may require additional attention

- **Regional performance**: By comparing the total sales amount for different regions, you can gain insight into which regions are performing well and which ones may require additional resources or attention

- **Product category performance**: By comparing the total sales amount for different product categories, you can gain insight into which product categories are performing well and which ones may require additional resources or attention

- **Total sales**: By summing up the total sales amount for all the regions, you can find out the total sales of the company

- **Average sales**: By averaging the total sales amount for all the regions, you can find out the average sales amount of the company

- **Comparison**: By comparing the values of different regions and product categories, you can find out which product is performing well in which region and which region is performing well in which product

By having these insights, you can make informed decisions on how to allocate resources, target marketing efforts, and improve sales performance.

Unpivoting data in SQL

Unpivoting data in SQL is the process of converting columns into rows. It is the inverse operation of the pivot function and is used to convert a table with a specific structure into a more flexible format.

The syntax for unpivoting data in SQL varies depending on the SQL DBMS being used. Here is an example of unpivot query syntax in SQL Server:

```
SELECT
sale_id,
attribute, value
FROM sales_data
UNPIVOT (value FOR attribute IN (product_category, region, sale_
amount)) AS u
```

In this example, the query uses the UNPIVOT function to convert the columns "product_category", "region", and "sale_amount" into rows, with a new column, "attribute", indicating the original column name, and a new column, "value", indicating the original value.

In MySQL and Oracle, you can use the UNION operator to achieve the same result:

```
SELECT
        sale_id,
        'product_category' as attribute,
        product_category as value FROM sales_data
UNION
SELECT sale_id,
        'region' as attribute,
        region as value FROM sales_data
UNION
SELECT sale_id,
        'sale_amount' as attribute,
        sale_amount as value
FROM sales_data
```

sale_id	attribute	value
1	product_category	Electronics
1	region	North
1	sale_amount	500
2	product_category	Clothing
2	region	South
2	sale_amount	300
3	product_category	Electronics
3	region	East
3	sale_amount	400
4	product_category	Furniture
4	region	West
4	sale_amount	800

Figure 6.12 – Query result

This unpivot operation can be useful when you need to perform analysis on specific columns or for data normalization. It can also make it easier to insert, update, or delete specific data in your table as it creates a more flexible format.

Analytical workflow – from SQL to business intelligence – transforming data into actionable insights

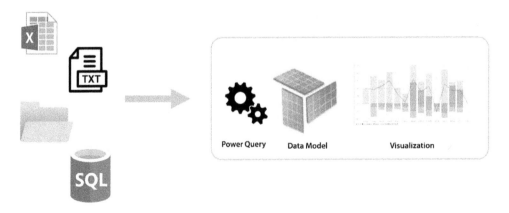

Figure 6.13 – SQL to BI workflow

The analytical workflow from SQL to **business intelligence (BI)** involves several steps that transform raw data stored in a database using SQL queries into actionable insights through BI tools. Here's a high-level overview of the process:

1. **Data extraction**: The first step is to extract the relevant data from the database using SQL. This involves writing queries to retrieve specific data tables, columns, and records that are needed for analysis. SQL's SELECT statement is commonly used for this purpose.

2. **Data transformation**: Data transformation plays a crucial role in the analytical process, as it prepares raw data for meaningful analysis and empowers organizations to derive actionable insights. SQL provides various transformation steps, including joins, union, pivoting, and aggregation, which are essential for effective data analysis. Here's an explanation of their importance:

   • **Joins**: Joins are used to combine data from multiple tables based on common columns. By linking related data, joins allow for comprehensive analysis across different dimensions. They help consolidate information and enable deeper exploration of relationships, such as connecting customer data with purchase history or merging sales data with product details. Joins provide a holistic view of the data, facilitating accurate and comprehensive analysis.

   • **Union**: The union operation combines rows from multiple tables or result sets into a single result set. This is beneficial when data needs to be merged vertically, such as appending records from multiple sources or consolidating data from different time periods. Union allows for the integration of disparate datasets, enabling comprehensive analysis across various data sources.

   • **Pivoting**: As we discussed earlier, pivoting is a transformation technique that rotates rows into columns, allowing for a different representation of data. It transforms tabular data from a long format (with many rows) into a wide format (with fewer rows and more columns). Pivoting is valuable when analyzing data with multiple dimensions or when comparing data across different attributes. It simplifies data analysis and enhances data visualization capabilities.

   • **Aggregation**: Aggregation combines multiple rows into a single summarized value, often using functions such as SUM, AVG, COUNT, MAX, or MIN. Aggregations provide insights into trends, patterns, and overall performance by consolidating data at different levels of granularity. They enable the calculation of **key performance indicators (KPIs)** and metrics, facilitating data-driven decision-making. Aggregations can be applied to various dimensions, such as time periods, regions, or product categories, to obtain valuable insights.

By performing these transformation steps in SQL, organizations can transform raw data into a structured and cohesive format that is suitable for analysis. Once the data is transformed, loading it into a BI tool for further analysis offers several advantages, as follows:

   • **Data loading**: After transforming the data, it needs to be loaded into a dedicated BI platform or tool for further analysis. This typically involves creating a data model or schema within the BI tool, which defines the relationships between various tables and dimensions. The loaded data is usually stored in a multidimensional or tabular format optimized for analytical processing.

- **Report and dashboard creation**: With the data loaded into the BI tool, reports and dashboards can be created. Reports are typically static representations of data, presenting specific metrics or visualizations. Dashboards, on the other hand, are interactive and allow users to explore data in real time, apply filters, drill down into details, and customize views. BI tools offer a variety of visualization options, such as charts, graphs, tables, and maps, to present data in a meaningful way.

- **Data analysis**: Users can analyze the data using the built-in analytical capabilities of the BI tool. This includes slicing and dicing data, applying filters, sorting, comparing metrics, creating calculated measures, and performing ad hoc queries. The goal is to gain insights and identify patterns, trends, and correlations within the data.

- **Data visualization**: Once the analysis is done, the results can be visualized using charts, graphs, and other visual representations. Effective data visualization helps users understand complex information quickly and supports data-driven decision-making. BI tools offer a wide range of visualization options, including bar charts, line charts, pie charts, and heatmaps.

- **Sharing and collaboration**: The insights gained from the data analysis can be shared with stakeholders and decision-makers. BI tools provide features to distribute reports, dashboards, and visualizations via email or scheduled notifications, or by embedding them into other applications or portals. Collaboration features allow users to annotate, comment on, and discuss the data with others, fostering a data-driven culture within the organization.

- **Monitoring and iteration**: BI is an iterative process. Users continuously monitor the performance of metrics and KPIs, identify areas for improvement, and refine their analysis. By collecting feedback, making adjustments, and exploring new questions, the analytical workflow evolves to meet changing business needs.

Overall, the analytical workflow from SQL to BI involves extracting, transforming, and loading data, followed by report and dashboard creation, data analysis, visualization, sharing, and continuous iteration. This process enables organizations to derive valuable insights from their data and make informed decisions.

Summary

This brings us to the end of this chapter. To summarize, pivoting data using SQL is a powerful technique for reorganizing and summarizing data in a more meaningful way. By using the `crosstab` or `pivot` function, you can convert columns into rows, effectively creating a new table with a different layout that makes it easier to perform calculations and generate meaningful reports. This can be particularly useful when working with data that has multiple dimensions, such as sales data by product category and region. The `pivot` function can also be used to generate dynamic columns based on unique values in one column and aggregate the other column values as per these unique values. Understanding and mastering the use of the `pivot` function in SQL is a valuable skill for any data analyst, data scientist, or data engineer working with large datasets.

Part 3:
SQL Subqueries, Aggregate
And Window Functions

This part includes the following chapters:

7

Subqueries and CTEs

This chapter explores the vital concepts of subqueries and **common table expressions (CTEs)** in SQL for data manipulation. We cover their usage within SELECT statements, including filtering, aggregation, and joining data. Advanced topics such as nested and correlated subqueries are discussed, along with the introduction of CTEs to break down complex transformations. Performance considerations and optimization techniques are emphasized. Practical examples demonstrate the application of subqueries and CTEs in real-world data-wrangling tasks, showcasing their versatility and power. By the end, you'll have a comprehensive understanding of leveraging these features for efficient data analysis and decision-making.

In this chapter, we will cover the following main topics:

- Introduction to subqueries
- Using subqueries in SELECT statements
- Nested subqueries
- Correlated subqueries
- Using subqueries in INSERT, UPDATE, and DELETE statements
- Common table expressions
- Performance considerations for subqueries and CTEs

Introduction to subqueries

A subquery is a query nested within another query, where the inner query is executed first and its result is used as input to the outer query. In other words, a subquery is a query within a query. Subqueries can be used to return data that is used in the main query as a condition to further restrict the data that is retrieved.

A subquery is a SELECT statement that is nested inside another SELECT, INSERT, UPDATE, or DELETE statement. The result of the subquery is used to modify the data retrieved by the main query.

Subqueries are used to retrieve data from multiple tables based on complex conditions and to aggregate and summarize data in a way that cannot be done with a single query.

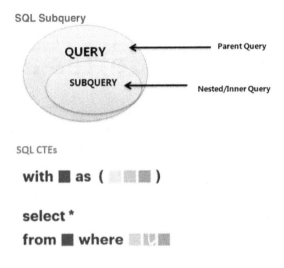

Figure 7.1 – Subquery

The syntax is as follows:

```
SELECT column_list
FROM table_name
WHERE column_name operator (SELECT column_name
                            FROM table_name
                            WHERE condition);
```

Here, column_list is a list of columns to be retrieved from the main query. table_name refers to the table from which the data is to be retrieved. The WHERE clause specifies the condition for the subquery and the main query. column_name is the name of the column used in the condition, and operator is a comparison operator, such as =, <, or >.

It's worth noting that subqueries can also be used in INSERT, UPDATE, and DELETE statements, with a similar syntax. The basic structure is to include the subquery within parentheses and use its result as a source for the main query.

There are two types of subqueries:

- Simple subqueries
- Correlated subqueries

Let's take a brief look at each one of them.

Simple subqueries

A simple subquery is an independent query that can be executed on its own and does not rely on the result of the main query. Let's say we have two tables: Customers and Orders. The Customers table contains customer information, while the Orders table contains order details. We want to find all customers who have placed an order:

```
SELECT * FROM Customers
WHERE CustomerID IN (SELECT CustomerID FROM Orders)
```

In this example, the (SELECT CustomerID FROM Orders) subquery retrieves all the customer IDs from the Orders table. The main query then uses this subquery to select all rows from the Customers table where the customer ID exists in the result of the subquery.

Correlated subqueries

A correlated subquery is a query that depends on the result of the main query. It executes for each row returned by the main query. Let's use the same Customers and Orders tables to find customers who have placed more than two orders:

```
SELECT * FROM Customers c
WHERE (SELECT COUNT(*) FROM Orders o WHERE o.CustomerID =
c.CustomerID) > 2
```

In this example, the (SELECT COUNT(*) FROM Orders o WHERE o.CustomerID = c.CustomerID) subquery counts the number of orders for each customer. The main query selects all rows from the Customers table and uses the correlated subquery to check whether the count of orders for each customer is greater than 2.

The correlated subquery is executed for each row in the main query's result set, comparing the customer ID of each row with the customer ID in the subquery.

Simple subqueries are independent and do not depend on the result of the main query. Correlated subqueries are dependent on the result of the main query and execute for each row returned by the main query. Subqueries can be used to perform complex data manipulations, such as calculating running totals, finding the maximum value in a group of records, and finding duplicates in a table.

Subqueries are a powerful tool for data manipulation in SQL and play an important role in data wrangling.

Using subqueries in SELECT statements

A subquery can be used in a SELECT statement to return data that will be used as a condition in the main query. The subquery runs first and returns a result set, which is then used by the main query to retrieve data.

The subquery can be placed in the SELECT clause to return a single value.

For example, a subquery in the SELECT clause could be used to calculate the average of a set of values and return it as a single value in the main query.

Example 1 – subquery in the SELECT clause

Consider a database table named orders with the following data:

order_id	customer_id	order_date
1	101	1/1/2022
2	102	2/1/2022
3	101	3/1/2022
4	103	4/1/2022

Figure 7.2 – orders table

To retrieve the order_date of the latest order made by each customer, the following query can be used:

```
SELECT customer_id, (SELECT MAX(order_date)
                     FROM orders
                     WHERE orders.customer_id = o.customer_id) AS
latest_order_date
FROM orders o
GROUP BY customer_id;
```

The (SELECT MAX(order_date) FROM orders WHERE orders.customer_id = o.customer_id) subquery retrieves the latest order date for each customer and returns it to the main query where it is used to restrict the result set to only the latest order made by each customer. The result of the preceding query will be as follows:

customer_id	latest_order_date
101	3/1/2020
102	2/1/2020
103	4/1/2020

Figure 7.3 – Resulting table

Using subqueries in FROM statements

A subquery can be used in the FROM clause to return a set of rows that can be used in the main query. The subquery acts as a derived table, which can be given an alias and referred to in the main query.

A subquery in the FROM clause is used to return a set of rows.

For example, consider a database table named orders with the following data:

order_id	customer_id	order_date
1	101	1/1/2020
2	102	2/1/2020
3	101	3/1/2020
4	103	4/1/2020

Figure 7.4 – orders table

Also consider another table, named customers, with the following data:

customer_id	customer_name
101	Jane Miller
102	Joesph Scott
103	Michael Jordan

Figure 7.5 – customers table

To retrieve the order_id, order_date, and customer_name for all orders made by customers, the following query can be used:

```
SELECT o.order_id, o.order_date, c.customer_name
FROM orders o
JOIN (SELECT customer_id, customer_name FROM customers) AS c
ON o.customer_id = c.customer_id;
```

In this example, the (SELECT customer_id, customer_name FROM customers) subquery returns customer_id and customer_name from the customers table and is given the alias c. The main query then joins the orders table with the derived table c on the customer_id to return the desired result set.

The result of the preceding query will be as follows:

order_id	order_date	customer_name
1	1/1/2020	Jane Miller
2	2/1/2020	Joseph Scott
3	3/1/2020	Jane Miller
4	4/1/2020	Michael Jordan

Figure 7.6 – Resulting table

Using subqueries in WHERE statements

A subquery is a SELECT statement that is nested within another SQL statement. It can be used in the WHERE clause of a main query to return a single value that is then used to filter the main query's results.

> **Note**
> A subquery should be added on the right side of the comparison operator to improve the performance of the query.

Here's an example.

Consider a table named orders with the following data:

order_id	customer_id	order_date	order_amount
1	101	1/1/2021	100
2	102	1/2/2021	200
3	103	1/3/2021	150
4	101	1/4/2021	50

Figure 7.7 – orders table

Now, we want to retrieve all orders for customers who have placed an order for an amount greater than the average order amount for all customers.

> **Note**
> This can be any operator, such as >, <, or =. The comparison operator can also be a multiple-row operator, such as IN, ANY, or ALL.

We can achieve this by using a subquery in the WHERE clause of the main query:

```
SELECT order_id, customer_id, order_date, order_amount
FROM orders
WHERE order_amount > (SELECT AVG(order_amount) FROM orders);
```

The subquery in this example returns the average order amount for all customers, which is then used to filter the main query's results. The main query returns only those orders where the order amount is greater than the average order amount. So, the result of this query will be as follows:

order_id	customer_id	order_date	order_amount
2	102	1/2/2021	200
3	103	1/3/2021	150

Figure 7.8 – Resulting table

Note that the subquery runs first and returns a single value, which is then used to filter the main query's results. This allows you to include complex logic and calculations in your WHERE clause, making it more flexible and powerful.

Use case scenarios

An interesting use case scenario for using subqueries in SQL is in HR analytics to analyze employee performance in a company.

Suppose we have a table named employees with the following columns: employee_id, employee_name, hire_date, and salary. We also have another table named performance_reviews with the employee_id, review_date, and rating columns.

employee_id	employee_name	hire_date	salary
1	Michael Harvey	1/1/2018	55000
2	Mike Smith	3/1/2019	60000
3	Jason Maynard	5/1/2017	65000
4	Henry S	7/1/2020	63000

employee_id	review_date	rating
1	12/1/2020	4
2	3/1/2021	5
3	12/1/2019	3
4	6/1/2022	4

Figure 7.9 – employees and performance_reviews tables

We want to retrieve the names of employees who were hired in the past five years and have received a performance review with a rating of 4 or higher.

Here's how we can use a subquery to achieve this:

```
SELECT
employees.
employee_name
FROM employees
WHERE hire_date >= DATE_SUB(CURDATE(), INTERVAL 5 YEAR)
AND employee_id IN (SELECT employee_id FROM performance_reviews WHERE
rating >= 4);
```

The first part of the WHERE clause filters the employees table to only include employees who were hired within the past five years:

```
hire_date >= DATE_SUB(CURDATE(), INTERVAL 5 YEAR)
```

The second part of the WHERE clause uses a subquery to identify employees who have received a performance review with a rating of 4 or higher:

```
AND employee_id IN (SELECT employee_id FROM performance_reviews WHERE
rating >= 4)
```

The subquery returns the employee IDs for employees who have received a performance review with a rating of 4 or higher. The main query then uses this list of IDs to filter the results and return only those employees who meet the specified criteria.

The final result of the query will be as follows:

employee_name
Michael Harvey
Mike Smith
Henry S

Figure 7.10 – Resulting table

These are the names of employees who were hired in the past five years and have received a performance review with a rating of 4 or higher.

This is just one example of how subqueries can be used to analyze data in a real-world scenario. By using subqueries, we can incorporate complex conditions and calculations into our queries to gain deeper insights into our data.

Here's another interesting industry example of how subqueries can be used in SQL, in the field of e-commerce.

Suppose an e-commerce company wants to identify its best-selling products in the past year. The company has two tables: `products` and `orders`. The `products` table contains information about each product, including its product ID, name, and category. The `orders` table contains information about each order, including the order date, product ID, and quantity sold.

product_id	name	category
1	iPhone 12	Electronics
2	MacBook Pro	Electronics
3	Adidas T-shirt	Clothing
4	Nike Sneakers	Shoes
5	Samsung TV	Electronics

Figure 7.11 – products table

order_date	product_id	quantity
12/1/2022	1	2
11/15/2022	2	1
10/20/2022	3	5
9/10/2022	4	3
8/1/2022	5	1
7/1/2022	1	1
6/15/2022	3	2

Figure 7.12 – orders table

Here's how the company could use a subquery to achieve this:

```
SELECT products.product_id, products.name, SUM(orders.quantity) AS
total_sold
FROM products
JOIN (
  SELECT product_id, SUM(quantity) AS quantity
  FROM orders
  WHERE order_date >= DATE_SUB(CURDATE(), INTERVAL 1 YEAR)
  GROUP BY product_id
) AS orders
ON products.product_id = orders.product_id
GROUP BY products.product_id
ORDER BY total_sold DESC
LIMIT 10;
```

The query aims to find the top 10 best-selling products of an e-commerce company in the past year. To achieve this, the query uses a subquery and a `JOIN` operation to combine data from two tables: `products` and `orders`.

The subquery in this query calculates the total quantity sold for each product in the past year:

```
SELECT product_id, SUM(quantity) AS quantity
FROM orders
WHERE order_date >= DATE_SUB(CURDATE(), INTERVAL 1 YEAR)
GROUP BY product_id
```

The subquery selects the `product_id` and `quantity` columns from the `orders` table. The `WHERE` clause filters the results to only include orders made in the past year. The `GROUP BY` clause aggregates the results by `product_id` to calculate the total quantity sold for each product. The result of the subquery is a table with two columns: `product_id` and `quantity`.

The main query then joins the result of the subquery with the `products` table:

```
SELECT products.product_id, products.name, SUM(orders.quantity) AS
total_sold
FROM products
JOIN (
    ... Subquery ...
) AS orders
ON products.product_id = orders.product_id
GROUP BY products.product_id
ORDER BY total_sold DESC
LIMIT 10;
```

The `JOIN` operation combines the data from the `products` and `orders` tables based on the `product_id` column. The `GROUP BY` clause aggregates the results by `product_id` to calculate the total quantity sold for each product. The `ORDER BY` clause sorts the results by the `total_sold` column in descending order to find the top 10 best-selling products. Finally, the `LIMIT` clause limits the results to the top 10 products.

name	total_quantity
MacBook Pro	1
Samsung TV	1
Nike Sneakers	3
iPhone 12	3
Adidas T-shirt	7

Figure 7.13 – Resulting table

A few of the key KPIs that stakeholders can get from this are the following:

- **Popular products**: The total quantity of each product sold can give an idea of which products are more popular among customers

- **Sales volume**: The total quantity of each product sold can give an insight into the overall sales volume of the business

- **Stock management**: The total quantity of each product sold can help in managing the stock levels and avoiding stock-outs or overstocking

- **Product demand**: The total quantity of each product sold can help in understanding the demand for each product, which can be used for future production planning

- **Customer preference**: The total quantity of each product sold can give an idea of customer preferences, which can be used for product development and marketing strategies

Nested subqueries

A nested subquery is a subquery that is contained within another subquery. It is used when a subquery result is used as input for another subquery. In other words, a nested subquery is a subquery inside another subquery.

For example, consider two tables, `employees` and `departments`, where the `employees` table contains the following data:

id	Name	salary	department_id
1	John	50000	1
2	Jane	55000	2
3	Alice	60000	2
4	Bob	65000	3

Figure 7.14 – employees table

The `departments` table contains the following data:

id	Name
1	IT
2	HR
3	Finance

Figure 7.15 – departments table

Now, suppose we want to find the name of the department with the highest average salary. To solve this, we can use a nested subquery, as follows:

```
SELECT name
FROM departments
WHERE (
   SELECT AVG(salary)
   FROM employees
   WHERE employees.department_id = departments.id
) = (
   SELECT MAX(AVG(salary))
   FROM employees
   GROUP BY department_id
);
```

The nested subquery inside the WHERE clause is used to calculate the average salary of employees in each department. The outer query then compares this average salary to the highest average salary calculated by the second subquery to find the department with the highest average salary.

To break down the nested subquery, it can be explained as follows:

- **First subquery**: (SELECT AVG(salary) FROM employees WHERE employees. department_id = departments.id) calculates the average salary of employees in each department by joining the employees and departments tables based on the department_id column

- **Second subquery**: (SELECT MAX(AVG(salary)) FROM employees GROUP BY department_id) calculates the highest average salary across all departments

- **Outer query**: This compares the average salary of each department, as calculated by the first subquery, to the highest average salary, as calculated by the second subquery, to find the department with the highest average salary

The resulting output of the query would look as follows:

Figure 7.16 – Resulting table

In this example, the nested subquery was used to find the name of the department with the highest average salary, which is the finance department.

Use case scenario

An interesting industry example of how nested subqueries can be used in SQL is in the retail industry. Let's consider a scenario where a retailer wants to find the products that have the highest profit margin in each category. The retailer has two tables: products and categories.

Id	name	price	cost	category_id
1	T-Shirt	25	10	1
2	Jeans	75	50	2
3	Sneakers	100	60	3
4	Hoodie	50	30	1

Figure 7.17 – products table

Id	name
1	Clothing
2	Footwear
3	Accessories

Figure 7.18 – categories table

The retailer can use the following nested subquery to find the product with the highest profit margin in each category:

```
SELECT
products.name,
(price - cost) as profit_margin,
categories.name as category
FROM products
JOIN categories
ON products.category_id = categories.id
WHERE (price - cost) = (
   SELECT MAX((price - cost))
   FROM products
   WHERE products.category_id = categories.id
);
```

In this example, the first subquery calculates the maximum profit margin for each category by joining the `products` and `categories` tables based on the `category_id` column. The outer query then compares the profit margin of each product to the maximum profit margin for its category to find the product with the highest profit margin in each category. The resulting output of the query would look as follows:

name	profit_margin	category
T-Shirt	15	Clothing
Jeans	25	Footwear
Sneakers	40	Accessories

Figure 7.19 – Resulting table

In this example, the nested subquery was used to find the product with the highest profit margin in each category, which is `T-Shirt` in `Clothing`, `Jeans` in `Footwear`, and `Sneakers` in `Accessories`.

A few KPIs that can be derived from this output are as follows:

- **Profit margin**: The profit margin is the difference between the price and cost of the product. This KPI shows how much profit the retailer is making on each product.

- **Product performance**: The name of the product with the highest profit margin in each category can be used to evaluate the performance of each product. This can help the retailer to identify the best-selling products and optimize their inventory accordingly.

- **Category performance**: The category name with the highest profit margin can be used to evaluate the performance of each category. This can help the retailer to understand which category is generating the highest profit and make strategic decisions based on that information.

- **Revenue**: The profit margin can be multiplied by the number of units sold to calculate the revenue generated by each product. This KPI can help the retailer to understand which products are contributing the most to their revenue.

Correlated subqueries

Earlier in this chapter, we briefly discussed the topic of correlated subqueries. Let's now take a deeper look at it. A correlated subquery is a type of subquery in SQL that depends on the values from the outer query. Unlike a regular subquery, which is executed first and then passed as a parameter to the outer query, a correlated subquery is executed for each row of the outer query and its result is used to filter the data for that specific row.

For example, consider a table named `Orders` that contains information about all the orders placed by customers. The table has the following columns: `OrderID`, `CustomerID`, `OrderDate`, and `TotalAmount`. You want to find out the total amount spent by each customer. You can use a correlated subquery to solve this problem.

OrderID	CustomerID	OrderDate	TotalAmount
1	1001	1/1/2021	100
2	1002	2/1/2021	150
3	1001	3/1/2021	200
4	1003	4/1/2021	175
5	1002	5/1/2021	125

Figure 7.20 – Orders table

Here's an example of what the query would look like:

```
SELECT CustomerID, SUM(TotalAmount) AS TotalSpent
FROM Orders o1
WHERE TotalAmount = (SELECT MAX(TotalAmount)
                     FROM Orders o2
                     WHERE o1.CustomerID = o2.CustomerID)
GROUP BY CustomerID;
```

CustomerID	TotalSpent
1001	300
1002	275
1003	175

Figure 7.21 – Resulting table

The correlated (SELECT MAX(TotalAmount) FROM Orders o2 WHERE o1.CustomerID = o2.CustomerID) subquery is executed for each row of the outer query and its result is used to filter the data for that specific row. In this example, the correlated subquery returns the maximum total amount spent by each customer, and the outer query uses this value to filter the data and returns only the rows where the customer's total amount spent is equal to the maximum amount spent. The result will be a table showing the total amount spent by each customer.

Correlated subqueries are useful when you want to compare the values in a row of a table with values in other rows of the same or a different table. They are often used when performing complex data analysis and when you want to extract information from a database in a flexible and dynamic way.

Use case scenario

An interesting use case of correlated subqueries in the industry is in the analysis of employee performance. For example, consider a company that has a database with two tables: Employees and Sales. The Employees table contains information about each employee, including their ID, name, and hire date. The Sales table contains information about each sale made by an employee, including the employee's ID, the sale date, and the sale amount.

EmployeeID	Name	HireDate
1	Michael Smith	1/1/2020
2	John Stevens	2/1/2020
3	Chris Paul	3/1/2020

Figure 7.22 – Employees table

EmployeeID	SaleDate	SaleAmount
1	1/2/2020	100
1	1/3/2020	200
1	2/1/2020	150
2	2/1/2020	200
2	3/1/2020	250
3	3/1/2020	300

Figure 7.23 – Sales table

To determine the average sales amount made by each employee in their first year of employment, the company can use a correlated subquery in the following way:

```
SELECT
  Employees.EmployeeID,
  Employees.Name,
  (SELECT AVG(SaleAmount)
  FROM Sales
  WHERE Sales.EmployeeID = Employees.EmployeeID
    AND Sales.SaleDate BETWEEN Employees.HireDate AND DATE_
ADD(Employees.HireDate, INTERVAL 1 YEAR)) AS FirstYearAvgSale
FROM Employees;
```

EmployeeID	Name	FirstYearAvgSale
1	Michael Smith	150
2	John Stevens	225
3	Chris Paul	300

Figure 7.24 – Resulting table

The subquery calculates the average sale amount made by each employee in their first year of employment, by taking the average of the `SaleAmount` column in the `Sales` table, filtered by the employee ID of each employee and the sale date being within the first year of their employment (between the hire date and one year after the hire date). The outer query then returns the employee's ID and name and the average sale amount, obtained from the subquery.

From the preceding result, the following **key performance indicators (KPIs)** can be derived:

- **Average sale amount**: This KPI shows the average sale amount made by each employee in their first year of employment

- **Employee performance**: This KPI allows stakeholders to compare the performance of employees and identify top performers

- **Sales trend**: This KPI shows the trend of sales for each employee and can help stakeholders to identify areas where improvement is needed

- **Sales growth**: This KPI shows the growth of sales made by each employee and helps stakeholders to identify areas where growth is needed

Using subqueries in INSERT, UPDATE, and DELETE statements

Subqueries can be used in `INSERT`, `UPDATE`, and `DELETE` statements in SQL to perform complex operations:

- `INSERT`: Subqueries can be used in an `INSERT` statement to retrieve data from a table and insert it into a new table. For example, you can insert data into a new table based on data retrieved from another table using a `SELECT` statement as a subquery:

```
INSERT INTO new_table (column1, column2, ...)
SELECT column1, column2, ...
FROM existing_table
WHERE some_condition;
```

- `UPDATE`: Subqueries can be used in an `UPDATE` statement to modify data based on data retrieved from another table. For example, you can update the values of a column in a table based on data retrieved from another table using a `SELECT` statement as a subquery:

```
UPDATE table1
SET column1 = (SELECT column2
               FROM table2
               WHERE some_condition)
WHERE some_condition;
```

- `DELETE`: Subqueries can be used in a `DELETE` statement to delete data based on data retrieved from another table. For example, you can delete rows from a table based on data retrieved from another table using a `SELECT` statement as a subquery:

```
DELETE FROM table1
WHERE column1 IN (SELECT column2
                  FROM table2
                  WHERE some_condition);
```

Subqueries are used in `INSERT`, `UPDATE`, and `DELETE` statements because they allow you to perform complex operations with multiple tables and conditions, without having to write complex join statements. They also allow you to encapsulate complex logic and reuse it in multiple statements, which can improve the maintainability and readability of your SQL code.

Use case scenario

Let's look at practical business examples for using subqueries in `INSERT`, `UPDATE`, and `DELETE` statements.

`INSERT`: Imagine your company tracks the sales made by your sales representatives. You have a table named `sales_representatives` that contains information about each representative, and a table named `sales` that contains information about each sale they make. You want to create a new table named `top_sales` that only contains information about the top 10 sales made by each representative. You can use a subquery in an `INSERT` statement to accomplish this:

```
INSERT INTO top_sales (representative_id, sale_amount, sale_date)
SELECT representative_id, sale_amount, sale_date
FROM sales
WHERE representative_id IN (SELECT representative_id
                           FROM sales
                           GROUP BY representative_id
                           ORDER BY SUM(sale_amount) DESC
                           LIMIT 10);
```

id	name
1	John
2	Sarah
3	Michael

Figure 7.25 – Sales representatives table

id	representative_id	Amount
1	1	1000
2	2	2000
3	1	1500
4	2	2500
5	3	3000
6	1	1700
7	2	2600
8	3	3200
9	1	1300
10	2	2200

Figure 7.26 - sales table

id	representative_id	Amount
1	1	1700
2	1	1500
3	1	1300
4	2	2600
5	2	2500
6	2	2200

Figure 7.27: top_sales table

> **Note**
>
> The `top_sales` table only contains the top 10 sales made by each representative (three sales per representative in this example). The amounts are in descending order.

UPDATE: Imagine your company tracks the salaries of your employees. You have a table named `employees` that contains information about each employee, including their current salary. You want to give a 10% raise to all employees who have been with the company for more than five years. You can use a subquery in an UPDATE statement to accomplish this:

```
UPDATE employees
SET salary = salary * 1.1
WHERE hire_date <= (CURRENT_DATE - INTERVAL 5 YEAR);
```

This is what the `employees` table might look like before the update:

id	name	hire_date	current_salary	years_at_company
1	John	1/1/2016	50000	7
2	Jane	1/1/2017	55000	6
3	Bob	1/1/2018	60000	5

Figure 7.28 – employees table

And here is what the table will look like after the update:

id	Name	hire_date	current_salary	years_at_company
1	John	1/1/2016	55000	7
2	Jane	1/1/2017	60500	6
3	Bob	1/1/2018	60000	5

Figure 7.29 – employees table post UPDATE statement

DELETE: Imagine your company tracks the inventory of your products. You have a table named `products` that contains information about each product, including its current stock level. You want to delete all products that have been discontinued and have 0 stock left. You can use a subquery in a DELETE statement to accomplish this:

```
DELETE FROM products
WHERE product_id IN (
  SELECT product_id
  FROM products
  WHERE discontinued = 'Yes' AND stock_level = 0
);
```

product_id	product_name	stock_level	discontinued
1	Product A	10	No
2	Product B	0	Yes
3	Product C	20	No
4	Product D	0	Yes

Figure 7.30 – products table before the DELETE statement

After the above DELETE statement is executed, the `products` table could look like this:

product_id	product_name	stock_level	discontinued
1	Product A	10	No
3	Product C	20	No

Figure 7.31 – products table after the DELETE statement

The preceding examples demonstrate how subqueries can be used in INSERT, UPDATE, and DELETE statements to perform complex operations on your data, without having to write complex join statements. They also allow you to encapsulate complex logic and reuse it in multiple statements, which can improve the maintainability and readability of your SQL code.

Consider a ride-sharing company that wants to keep track of the total distance traveled by each driver, as well as the total revenue generated from their rides. The company has two tables: `drivers` and `rides`. The `drivers` table contains information about each driver, including their name and driver ID, and the `rides` table contains information about each ride, including the driver ID, distance traveled, and fare.

The company could use subqueries in an INSERT, UPDATE, and DELETE statement to update the `drivers` table with the total distance traveled and revenue generated by each driver. For example, the following query would insert a new row into the `drivers` table for each unique driver ID, with the total distance traveled and revenue generated:

```
INSERT INTO drivers (driver_id, total_distance, total_revenue)
SELECT driver_id, SUM(distance), SUM(fare)
FROM rides
GROUP BY driver_id;
```

driver_id	name
1	John
2	Jane
3	Bob

Figure 7.32 – drivers table

driver_id	distance_traveled	fare
1	10	20
1	5	15
2	15	25
3	12	22

Figure 7.33 – rides table

After the INSERT statement, the new driver table would look something like this:

driver_id	sum_of_distance	sum_of_fare
1	15	35
2	15	25
3	12	22

Figure 7.34 – Resulting table

Similarly, the following query would update the drivers table with the updated total distance traveled and revenue generated by each driver, based on the rides in the rides table:

```
UPDATE drivers SET total_distance = (
SELECT SUM(distance)
FROM rides
WHERE drivers.driver_id = rides.driver_id
GROUP BY driver_id
), total_revenue = (
SELECT SUM(fare)
FROM rides
WHERE drivers.driver_id = rides.driver_id
GROUP BY driver_id
```

```
);
```

The following query would delete a row from the `drivers` table if the corresponding driver no longer has any rides:

```
DELETE FROM drivers
WHERE driver_id NOT IN (
SELECT driver_id
FROM rides
GROUP BY driver_id
);
```

These queries demonstrate how subqueries can be used in `INSERT`, `UPDATE`, and `DELETE` statements to efficiently manage and update data in a relational database, in the context of a ride-sharing company.

The `INSERT`, `UPDATE`, and `DELETE` statements can be used to manage and update the data in these tables, which can then be used to calculate various KPIs. Here are some examples:

- `INSERT` statements can be used to add new trip data to the `trips` table, which can then be used to calculate the number of trips taken by drivers, the average trip duration, and the total revenue generated by the company

- `UPDATE` statements can be used to modify existing data in the trips table, such as updating the fare amount for a trip, which can then be used to recalculate the total revenue generated by the company

- `DELETE` statements can be used to remove data from the `trips` table, such as when a trip is canceled, which can then be used to recalculate the number of trips taken by drivers and the average trip duration

By using these statements to manage and update the data in the database, the KPIs can be generated in real time, allowing the ride-sharing company to monitor its performance and make data-driven decisions.

Managing and maintaining subqueries

Subqueries can become unreadable quickly in SQL code due to several factors:

- **Nesting**: When subqueries are nested deeply within each other, it can make the code difficult to follow and understand. As the level of nesting increases, it becomes harder to keep track of the logic and relationships between the subqueries.

- **Lack of clarity**: Subqueries that are not properly structured or lack clear aliases can make the code confusing. It may be challenging to determine the purpose and intended result of the subquery.

- **Complex logic**: Subqueries that involve complex logic, multiple conditions, or intricate joins can quickly become convoluted. The more complex the logic, the harder it is to comprehend and debug.

- **Inefficient use of subqueries**: Subqueries can also become unreadable if they are used unnecessarily or inefficiently. Using subqueries when simpler alternatives exist can make the code unnecessarily complex and difficult to follow.

To improve the readability of subqueries, it is essential to follow good coding practices:

- **Use descriptive aliases**: Assign meaningful aliases to subqueries to enhance readability and make it easier to understand their purpose.

- **Break down complex logic**: If a subquery involves complex logic, consider breaking it down into smaller, more manageable parts. Use temporary tables or CTEs to simplify the code and improve readability.

- **Limit nesting levels**: Avoid excessive nesting of subqueries. If possible, refactor the code to use joins or other SQL constructs instead.

- **Comment the code**: Add comments to explain the purpose and logic of the subqueries, especially if they involve complex operations. Clear comments can help other developers (and your future self) understand the code more easily.

By following these practices, you can enhance the readability of subqueries and make your SQL code more maintainable and understandable.

Common table expressions

CTEs in SQL are a convenient way to temporarily simplify complex queries by breaking them down into smaller, more manageable parts. A CTE is defined within a SELECT, INSERT, UPDATE, or DELETE statement and can be used to reference a temporary, named result set within the scope of a single query.

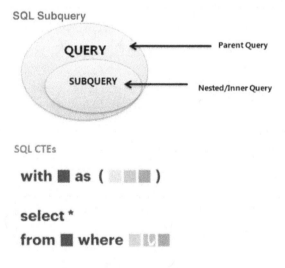

Figure 7.35 – SQL CTE expressions

A CTE is defined using the WITH clause, followed by a SELECT statement that specifies the columns and data for the temporary result set. The CTE is then referenced in the main query using its name, just like a table in the database.

Here is an example of how a CTE can be used in a SQL query:

```
WITH TripInformation AS (
  SELECT
    drivers.name AS driver_name,
    trips.start_time AS trip_start_time,
    trips.end_time AS trip_end_time,
    trips.fare AS trip_fare
  FROM
    drivers
    JOIN trips ON drivers.id = trips.driver_id
)
SELECT
  driver_name,
  SUM(trip_fare) AS total_fare
FROM
  TripInformation
GROUP BY
  driver_name
ORDER BY
  total_fare DESC;
```

In this example, a CTE named TripInformation is defined to represent the result of a query that combines data from the drivers and trips tables. The main query then references the CTE to calculate the total fare earned by each driver.

CTEs can make SQL queries more readable and easier to maintain by reducing the complexity of the main query and providing a named reference to intermediate results.

Use case scenario

Here is an example of how a CTE can be used in a real-world scenario in the hospitality industry.

Let's say a hotel wants to find the total number of rooms occupied by guests for each month of the year. Here's an example query that uses a CTE to accomplish this:

```
WITH MonthlyOccupancy AS (
  SELECT
    date_trunc('month', check_in_date) AS month,
    count(*) AS rooms_occupied
  FROM
```

```
    reservations
  WHERE
    check_in_date >= date '2021-01-01'
    AND check_in_date < date '2022-01-01'
  GROUP BY
    month
)
SELECT
  to_char(month, 'Mon YYYY') AS month,
  rooms_occupied
FROM
  MonthlyOccupancy
ORDER BY
  month;
```

In this example, a CTE named `MonthlyOccupancy` is defined to represent the result of a query that calculates the number of rooms occupied by guests for each month of the year. The main query then references the CTE to present the results in a readable format, with each month represented as a string, such as `Jan 2022`.

check_in_date
1/1/2022
1/3/2022
2/1/2022
2/2/2022
3/1/2022

Figure 7.36 – reservations table

month	rooms_occupied
1/1/2022	2
2/1/2022	2
3/1/2022	1

Figure 7.37 – MonthlyOccupancy table

Month	rooms_occupied
Jan-2022	2
Feb-2022	2
Mar-2022	1

Figure 7.38 – Resulting table

By using a CTE in this query, we can simplify the main query and make it more readable, as well as reuse the intermediate results of the CTE in other parts of the application.

Here's an interesting scenario for using CTEs in SQL data wrangling in the food delivery industry.

A food delivery company wants to analyze their delivery times to improve their efficiency and customer satisfaction. They have a database with information about all their delivery orders and want to find the average delivery time for each hour of the day.

The data analyst at the food delivery company creates a query that uses a CTE to find the average delivery time for each hour of the day. The CTE first groups the delivery orders by the hour they were placed and then calculates the average delivery time for each group. The main query references the CTE and presents the results in a readable format, with each hour represented as a string, such as 1 PM.

order_time	delivery_time
1/1/2022 11:00	30 minutes
1/1/2022 12:00	20 minutes
1/1/2022 18:00	45 minutes
1/2/2022 11:00	25 minutes
1/2/2022 18:00	50 minutes

Figure 7.39 – Orders table

order_hour	avg_delivery_time
11:00 AM	27.5 minutes
12:00 PM	20 minutes
6:00 PM	47.5 minutes

Figure 7.40 – CTE Table

Let's examine the following CTE query to see how it can be used:

```
WITH AvgDeliveryTime AS (
    SELECT
        DATEPART(HOUR, order_time) AS order_hour,
        AVG(delivery_time) AS avg_delivery_time
    FROM DeliveryOrders
    GROUP BY DATEPART(HOUR, order_time)
)
SELECT
    CAST(order_hour AS VARCHAR(2)) + ' PM' AS order_hour,
    avg_delivery_time
FROM AvgDeliveryTime;
```

order_hour	avg_delivery_time
11:00 AM	27.5 minutes
12:00 PM	20 minutes
6:00 PM	47.5 minutes

Figure 7.41 – Orders table

The results of the query show that the average delivery time was slowest during the hours of 6 P.M. to 8 P.M., when many people were ordering food for dinner. The delivery times were quickest during the hours of 11 A.M. to 1 P.M., when many people were ordering lunch.

Based on this data, the food delivery company decides to increase their staffing levels during the hours of 6 P.M. to 8 P.M. to improve their delivery times during the dinner rush. They also decide to focus on reducing their delivery times during the lunch hour to increase customer satisfaction.

By using a CTE in their analysis, the food delivery company was able to make data-driven decisions to improve their delivery times and better understand their performance.

Here are some additional KPIs that can be calculated using the preceding data:

- **Delivery time distribution**: This KPI shows the distribution of delivery times throughout the day. The distribution can be visualized using a histogram or a bar chart.

- **Delivery time variability**: This KPI measures the variability of delivery times. It can be calculated using the standard deviation of the delivery times. The lower the standard deviation, the more consistent the delivery times are.

- **Delivery time by location**: This KPI shows the average delivery time for each location served by the food delivery company. This information can be used to identify which locations are taking longer to deliver to and why.

- **Delivery time by type of food**: This KPI shows the average delivery time for different types of food. This information can be used to identify which types of food take longer to prepare and deliver and why.

- **Delivery time by day of the week**: This KPI shows the average delivery time for each day of the week. This information can be used to identify which days are busier and why.

By calculating these KPIs, the food delivery company can get a better understanding of their delivery times and identify areas for improvement.

Performance considerations for subqueries and CTEs

Subqueries and CTEs are powerful features in SQL that can simplify complex data-wrangling tasks. However, they can have an impact on the performance of your queries. Here are some performance considerations to keep in mind when using subqueries and CTEs:

- **Indexing**: When using subqueries and CTEs, it's important to have appropriate indexes on the relevant columns in the underlying tables. Otherwise, the query might perform slowly.

- **Data size**: The size of the data being processed can greatly affect the performance of subqueries and CTEs. Larger datasets require more processing power, which can slow down the query.

- **Query optimization**: The SQL engine optimizes the query execution plan based on various factors, such as the indexes, statistics, and query structure. The query optimizer may choose to process the subquery or CTE in a different way than you expect, which can impact the query performance.

- **Nested subqueries**: Using nested subqueries can be computationally expensive and may result in slow query performance and becomes unreadable. Consider using a JOIN instead of a nested subquery whenever possible.

- **CTE materialization**: When a CTE is used in a query, the SQL engine creates a materialized version of the CTE's result set that is stored in memory. A materialized version, in the context of CTEs, refers to the actual result set of a CTE that is stored in memory or disk as a temporary table-like structure. This materialized version is created when the CTE is referenced in a query and serves as a reusable data source for subsequent operations within that query. Storing the result set as a materialized version can improve performance by avoiding redundant computations and allowing the data to be accessed more efficiently. However, if the materialized version is large, it can consume a significant amount of memory or disk space, potentially affecting query performance.

- **Temporary table versus materialized view of CTEs**: A temp table, also known as a temporary table, is a table that is created and used temporarily within a session or a specific scope. Temp tables are used to store intermediate or temporary results during a specific operation or query. They are created with a specific structure, similar to regular database tables, and can be used for data manipulation and storage. Temp tables are typically used when there is a need to store and manipulate a significant amount of data temporarily.

  In summary, a temp table is a temporary table used for temporary data storage within a session or scope, while a materialized view is a precomputed and stored result set that improves query performance by allowing faster data access. Temp tables are used for temporary data manipulation, while materialized views are used for caching and optimizing frequently executed queries.

- **Query complexity**: Subqueries and CTEs can make the query more complex and harder to understand, which can make it difficult to optimize the query for performance.

In summary, subqueries and CTEs are useful features in SQL, but they should be used judiciously to ensure the best query performance. It's important to keep the data size, query optimization, nested subqueries, CTE materialization, and query complexity in mind when designing your queries.

Subquery versus CTEs

Subqueries and CTEs are two ways of retrieving data from multiple tables in SQL. However, there are a few key differences between them:

- **Definition**: A subquery is a query within another query, while a CTE is a temporary result set that can be referred to within a `SELECT`, `INSERT`, `UPDATE`, or `DELETE` statement.

- **Readability**: CTEs are generally easier to read and understand, as they provide a clear, named reference to the intermediate results being used in a query. In contrast, subqueries can become difficult to read, especially when nested within multiple levels of a query.

- **Reusability**: CTEs can be reused multiple times within a query, which makes the code more concise and maintainable. Subqueries, on the other hand, are limited to a single use within a query and must be repeated for each use.

- **Performance**: The performance of subqueries and CTEs can vary depending on the complexity of the query and the data being retrieved. However, CTEs can provide better performance, as they allow the optimizer to reuse intermediate results and avoid redundant computation.

Overall, both subqueries and CTEs can be useful for retrieving data from multiple tables in SQL, but CTEs are generally preferred for their improved readability, reusability, and performance.

Summary

This brings us to the end of this chapter. So far, we have learned how SQL subqueries and CTEs are used to manipulate and retrieve data from multiple tables. Subqueries are used within another SQL statement and return a single value or a set of values to be used in the main query. CTEs provide a named, temporary result set that can be referenced within a `SELECT`, `INSERT`, `UPDATE`, or `DELETE` statement and are used to simplify complex SQL statements. To effectively use subqueries and CTEs, it's important to understand the difference between them and how they can impact query performance. In conclusion, by mastering subqueries and CTEs, you can become a more effective and efficient SQL data wrangler.

In the next chapter, we will learn about SQL aggregate functions and understand how they can be used in the process of statistical analysis and data wrangling.

Aggregate Functions

In this chapter, we will explore the topic of data wrangling using aggregate functions in SQL. We will begin delving into the various aggregate functions available in SQL, such as COUNT, SUM, AVG, MIN, and MAX, and show how they can be used to summarize and analyze data. We will also cover advanced aggregate functions such as COUNT DISTINCT. Throughout the chapter, we will provide real-world examples to illustrate the concepts discussed. By the end of this chapter, you will have a solid understanding of how to effectively use aggregate functions in SQL for data-wrangling tasks.

In this chapter, we will cover the following main topics:

- Overview of aggregate functions in SQL
- Advanced aggregate functions
- Using GROUP BY clauses to group data and apply aggregate functions

Overview of aggregate functions in SQL

Aggregate functions in SQL are used to summarize and analyze data in a table or a query's result set. These functions allow us to perform calculations on multiple rows of data and return a single value.

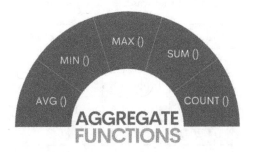

Figure 8.1 – Aggregate functions

The most common aggregate functions include the following:

- COUNT: This function returns the number of rows in a table or a result set that match a specific condition.

- SUM: This function returns the sum of all values in a specific column. It is commonly used to calculate the total value of a numeric column, such as total sales or total salary.

- AVG: This function returns the average value of a specific column. It is commonly used to calculate the average value of a numeric column, such as average salary or average price.

- MIN: This function returns the minimum value of a specific column. It is commonly used to determine the lowest value in a column, such as the lowest salary or the earliest date.

- MAX: This function returns the maximum value of a specific column. It is commonly used to determine the highest value in a column, such as the highest salary or the latest date.

In addition to these common aggregate functions, there are other advanced functions such as this one:

- COUNT DISTINCT: This function is used to count the number of distinct values in a column. This function can also be used to count the number of unique values in a column.

When using aggregate functions, it is important to group the data by one or more columns using a GROUP BY clause. This allows the aggregate functions to be performed on each group of data rather than on the entire table or result set.

Using GROUP BY

The GROUP BY clause in SQL is used to group rows in a result set based on one or more columns. This is often used in conjunction with aggregate functions such as SUM, COUNT, AVG, MIN, and MAX to perform calculations on the grouped data.

For example, let's say we have a table called sales that contains information about sales transactions, including the date of the sale, the product that was sold, and the sales revenue.

date	product	revenue
1/1/2021	Product A	100
1/2/2021	Product B	200
1/3/2021	Product A	150
1/4/2021	Product B	300
1/5/2021	Product C	250

Figure 8.2 – Sales table

We can use the GROUP BY clause to group all the rows in the table by the product that was sold and then use the SUM function to calculate the total revenue for each product. The resulting query would look something like this:

```
SELECT
product,
SUM(revenue)as total_revenue
FROM sales
GROUP BY product;
```

This query would return a result set with one row for each unique product in the table, with the product name in the first column and the total revenue for that product in the second column.

product	total_revenue
Product A	250
Product B	500
Product C	250

Figure 8.3 – Query result

We can also group by multiple columns, for example, like in the following:

```
SELECT
product,
date,
SUM(revenue)as total_revenue
FROM sales
GROUP BY product, date;
```

This query would return a result set with one row for each unique combination of product and date in the table, with the product name, date, and total revenue for that product in the respective columns.

Basically, the query would give total_revenue for each product and date.

It's also worth noting that in addition to aggregate functions, we can also use the GROUP BY clause with other SQL clauses such as ORDER BY and HAVING to further manipulate the grouped data.

Let's consider a table called Sales with the following columns: Product, Category, and Revenue. We want to calculate the total revenue, maximum revenue, and minimum revenue for each category.

Product	Category	Revenue
A	X	100
B	X	150
C	Y	200
D	Y	300
E	Y	250

Figure 8.4 – Sales table

Here's the SQL code to calculate the total revenue, maximum revenue, and minimum revenue for each category:

```
SELECT Category, SUM(Revenue) AS TotalRevenue, MAX(Revenue) AS
MaxRevenue, MIN(Revenue) AS MinRevenue
FROM Sales
GROUP BY Category;
```

Category	TotalRevenue	MaxRevenue	MinRevenue
X	250	150	100
Y	750	300	200

Figure 8.5 – Query result

In the preceding example, the GROUP BY clause is used to group the rows by the Category column. The aggregate functions, namely SUM, MAX, and MIN, are then applied to the Revenue column within each category.

The result set displays the total revenue, maximum revenue, and minimum revenue for each category. In this case, the X category has a total revenue of 250, with a maximum revenue of 150 and a minimum revenue of 100. The Y category has a total revenue of 750, with a maximum revenue of 300 and a minimum revenue of 200.

In summary, the GROUP BY clause in SQL allows us to group rows in a result set based on one or more columns and then use aggregate functions to perform calculations on the grouped data. This is a powerful tool for analyzing and summarizing large sets of data.

Case scenario

An interesting use case scenario for the GROUP BY clause in SQL could be analyzing survey data for a market research company. The company may have a table of survey responses, with columns for the respondent's age, gender, income, and overall satisfaction rating.

To understand the demographics of the respondents and their satisfaction levels, the market research company might use GROUP BY to group the responses by age and gender and then use aggregate functions such as COUNT() and AVG() to calculate the total number of respondents for each group and their average satisfaction rating.

respondent_id	age	gender	income	satisfaction_rating
1	25	Male	50000	4
2	30	Female	55000	5
3	35	Male	60000	3
4	40	Female	65000	4
5	45	Male	70000	4
6	50	Female	75000	5

Figure 8.6 – survey_responses table

For example, the following SQL query would group responses by age and gender and calculate the total number of respondents and average satisfaction rating for each group:

```
SELECT age,
  gender,
  COUNT(*) as num_respondents,
  AVG(satisfaction_rating) as avg_satisfaction
FROM survey_responses
GROUP BY age, gender
ORDER BY avg_satisfaction DESC;
```

This query would return a result set with one row for each unique combination of age and gender in the survey_responses table, showing the total number of respondents and average satisfaction rating for that group.

age	gender	num_respondents	avg_satisfaction
50	Female	1	5
30	Female	1	5
25	Male	1	4
40	Female	1	4
45	Male	1	4
35	Male	1	3

Figure 8.7 – Query result

The market research company can use this information to identify which demographic groups are the most satisfied with their products or services and target their marketing efforts accordingly.

So, let's begin by learning about these different aggregate functions and understanding how they can be used to summarize and analyze data effectively.

COUNT()

The COUNT() function in SQL is used to return the number of rows in a table or a result set that matches a specific condition. It is commonly used to count the number of records in a table or to count the number of occurrences of a specific value in a column.

The basic syntax for using the COUNT() function is as follows:

```
SELECT COUNT(column_name) FROM table_name;
```

In this syntax, column_name is the name of the column you want to count the rows of, and table_name is the name of the table you want to retrieve the data from.

You can also use the COUNT() function with the DISTINCT keyword to return the number of unique values in a column:

```
SELECT COUNT(DISTINCT column_name) FROM table_name;
```

In addition to counting the rows in a table or a result set, the COUNT() function can also be used with other SQL commands such as WHERE, GROUP BY, and HAVING to filter and group the data before counting it.

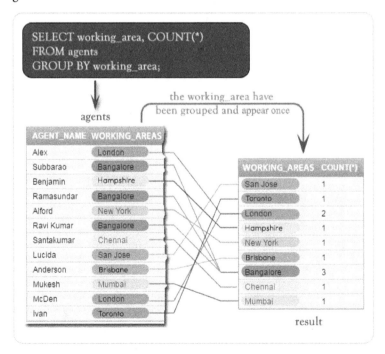

Figure 8.8 – Pictorial representation of SQL COUNT() and GROUP BY

Let's say we have an `orders` table and we want to count the total number of orders. We can write the following query:

```
SELECT COUNT(*) FROM orders;
```

In this example, the `COUNT()` function is used to count the total number of rows in the `orders` table. `*` is used as a wildcard to select all the columns. This is used when we don't have a specific column to count.

The result of this query would be a single value representing the total number of rows in the `orders` table.

Case scenario

The `COUNT()` function in SQL can be used by a retail company that wants to track the number of products sold per day. The company has a database table called `sales` that contains information about each product sale, including the date of the sale and the product ID.

sale_id	product_id	sale_date
1	101	1/24/2023
2	102	1/24/2023
3	103	1/24/2023
4	101	1/25/2023
5	104	1/25/2023
6	105	1/25/2023
7	101	1/26/2023
8	103	1/26/2023
9	105	1/26/2023
10	101	1/27/2023
11	102	1/27/2023
12	103	1/27/2023
13	104	1/27/2023
14	105	1/27/2023

Figure 8.9 – sales table

To find the number of products sold per day, the company would use the COUNT() function in a SQL query that groups the data by date and counts the number of rows in each group. The query would look something like this:

```
SELECT  sale_date,
   COUNT(*) as count_products
FROM sales
GROUP BY date;
```

This query would return a result set that shows the number of products sold each day, allowing the company to track sales trends over time and make informed business decisions.

sale_date	count_products
1/24/2023	3
1/25/2023	3
1/26/2023	3
1/27/2023	5

Figure 8.10 – Query result

An interesting example of using the COUNT() function in SQL could be in the social media industry, where a company wants to analyze the engagement of its users. The company may have a table that contains user information, including the user_id, registration_date, and the number_of_posts.

user_id	post_id	post_text	post_date	registration_date
1	1001	Hello	1/1/2021	1/1/2020
1	1002	World	1/2/2021	1/5/2020
2	2001	Greetings	1/3/2021	2/1/2020
2	2002	Everyone	1/4/2021	3/1/2020
3	3001	Hi	1/5/2021	3/2/2020
4	4001	Bye	1/6/2021	3/3/2020

Figure 8.11 – Sample posts table from a larger posts table

In this scenario, we can use the COUNT() function to calculate the number of posts each user has made on the platform. This information can be used to identify which users are the most active and which ones may be less engaged.

Here's an example of how this might be done:

```
SELECT
    user_id,
    COUNT(user_id) as number_of_posts,
    DATEDIFF(NOW(), registration_date) as days_since_registration
FROM posts
GROUP BY user_id
ORDER BY number_of_posts DESC
```

This query will give us the number of posts each user has made and the days since the user registration. It will group the data by `user_id` and order the results by number of posts in descending order, so we can see the most active users at the top.

user_id	number_of_posts	days_since_registration
1	150	30
2	70	45
3	30	120
4	8	150
5	35	80

Figure 8.12 – Query result

There are a number of insights that can be generated through the use of the COUNT() function, as shown in the preceding example. Here are some examples:

- **Identifying the most active users**: By counting the number of posts each user has made, it is possible to identify which users are the most active on the platform. This information can be used to target marketing and engagement efforts to these users.

- **Identifying patterns of engagement**: By grouping the data by user and ordering the results by the number of posts, it may be possible to identify patterns of engagement among users. For example, you may notice that users who have been on the platform for a longer period of time tend to be more active than newer users.

- **Identifying users who are at risk of becoming inactive**: By combining the number of posts with the days since registration, it is possible to identify users who have been active in the past but have become less active over time. This information can be used to target re-engagement efforts for these users.

SUM()

The SUM() function is used to calculate the sum of a selected column in a table. It takes the name of a column as an argument and returns the sum of all the values in that column. The basic syntax for using the SUM() function is as follows:

```
SELECT
SUM(column_name)
FROM table_name;
```

SUM(column_name) is the SUM function that calculates the sum of the values in the specified column.

order_id	customer_name	product	price
1	John Smith	T-shirt	20
2	Jane Doe	Hoodie	35
3	Bob Johnson	Hat	12.5
4	Sarah Lee	Jacket	60
5	Tom Wilson	Pants	25

Figure 8.13 – Orders table

For example, let's say you have a table named orders with the following columns: order_id, customer_name, product, and price. To calculate the total value of all the orders in the table, you would use the following query:

```
SELECT
SUM(price)
FROM orders;
```

This query would return the sum of all the values in the price column of the orders table, which would be the total value of all the orders. So, in this case, the total value of all the orders in the table is $152.50.

order_id	customer_name	product	price
1	John Smith	T-shirt	20
2	Jane Doe	Hoodie	35
3	Bob Johnson	Hat	12.5
4	Sarah Lee	Jacket	60
5	Tom Wilson	Pants	25
6	Jack Smith	T-shirt	22.5
7	Jill Doe	Hat	15
8	Mike Johnson	Jacket	50
9	Amy Lee	T-shirt	18

Figure 8.14 – Orders table

You can also use the GROUP BY clause to group the data by a certain column and then use the SUM function to get the sum of grouped data, as seen in *Figure 8.14*:

```
SELECT product,
    SUM(price)
FROM orders
GROUP BY product;
```

This query will group the data by product and return the sum of the price for each product.

product	SUM(price)
Hat	27.5
Hoodie	35
Jacket	110
Pants	25
T-shirt	60.5

Figure 8.15 – Result set

As you can see, the query has grouped the data by each unique value in the product column and calculated the sum of the price column for each group. For example, there are three orders for T-shirts in the table, and the sum of their prices is $60.50.

Case scenario

An interesting case scenario for using the SUM() function in SQL is analyzing sales data for a retail company. The company has a database with several tables containing information about its products, customers, and orders. They want to use SQL to analyze their sales data and find out which products are the most popular and generate the most revenue.

order_id	customer_name	product_name	price
1	Alice	T-Shirt	49.99
2	Bob	Hoodie	99.99
3	Charlie	Socks	24.99
4	Dave	Shoes	199.99
5	Eve	Hat	39.99
6	Alice	Hoodie	99.99
7	Bob	Socks	24.99
8	Charlie	Shoes	199.99
9	Dave	T-Shirt	49.99

Figure 8.16 – Orders table

To find the total sales for each product in this dataset without using a JOIN statement, we can group the data by the product_name column and calculate the total sales for each product by multiplying the price and quantity columns together and summing the results using the SUM() function. Here's the SQL query for that:

```
SELECT product_name,
   SUM(price) as total_sales
FROM orders
GROUP BY product_name
ORDER BY total_sales DESC;
```

This query will group the data by the product_name column and calculate the total sales for each product by summing the price column using the SUM() function. The results will be ordered in descending order of total sales. This will allow the company to see which products have the highest total sales and are the most popular among customers.

product_name	total_sales
Shoes	399.98
Hoodie	199.98
T-Shirt	99.98
Socks	49.98
Hat	39.99

Figure 8.17 – Query result

As you can see, the query has calculated the total sales for each product by summing the `price` column and grouping the data by the `product_name` column. Additionally, the company can use different filters, such as date range and location, to analyze the data more granularly. For example, if the company wants to know the revenue for a specific month for a specific location, it can use the following query:

order_id	customer_name	product_name	price	quantity	location	order_date
1	Alice	T-Shirt	49.99	2	New York	1/1/2022
2	Bob	Hoodie	99.99	1	New York	1/1/2022
3	Charlie	Socks	24.99	3	New York	1/2/2022
4	Dave	Shoes	199.99	1	Los Angeles	1/2/2022
5	Eve	Hat	39.99	2	Los Angeles	1/3/2022
6	Alice	Hoodie	99.99	1	New York	1/4/2022
7	Bob	Socks	24.99	1	Los Angeles	1/5/2022
8	Charlie	Shoes	199.99	2	Los Angeles	1/5/2022
9	Dave	T-Shirt	49.99	3	New York	1/6/2022

Figure 8.18 – Orders table

To calculate the revenue for a specific month and location, we can modify the previous SQL query by adding WHERE conditions for the `location` column and `order_date` column. Here's an example query that calculates the revenue for the month of January 2022 in New York:

```
SELECT product_name,
    SUM(price * quantity) as revenue
FROM orders
WHERE location = 'New York' AND order_date >= '2022-01-01' AND order_
date < '2022-02-01'
```

```
GROUP BY product_name
ORDER BY revenue DESC;
```

This query would give the company the revenue for each product in New York during the month of January 2022.

product_name	revenue
T-Shirt	299.97
Hoodie	199.98
Socks	74.97

Figure 8.19 – Query result

Overall, the SUM() function in SQL is a versatile and essential tool for analyzing and understanding sales data and can be used in many different ways to help a company make informed decisions about its products and business strategies.

An interesting example of how the SUM() function can be used in SQL is in the financial industry. A financial institution may have a database of financial transactions, with each record containing the account number, the transaction type (debit or credit), and the amount. The SUM() function can be used to calculate the total balance for an account.

transaction_id	account_number	transaction_type	amount
1	824567982	Debit	500
2	824567982	Credit	1000
3	824567982	Debit	750
4	824567982	Credit	200

Figure 8.20 – Transaction table

The SQL query might look like this:

```
SELECT
SUM(CASE WHEN transaction_type = 'credit' THEN amount ELSE
-       amount END) AS balance
FROM transactions
WHERE account_number = '824567982';
```

This query would calculate the total balance for account number '824567982' by summing up all the credit transactions and subtracting all the debit transactions.

The resulting value would be stored in the `balance` column:

```
1000.00 - 500.00 + (-750.00) + 200.00 = -50.00
```

balance
-50

Figure 8.21 – Query result

This information can be used to keep track of a customer's account balance, to ensure that there are enough funds before making a transaction, and to make decisions about loans, credit limits, and interest rates. Additionally, this metric can be used to monitor account activity and detect any suspicious or fraudulent activity.

Another important metric that can be derived from this query is the credit and debit transaction amount by using the following query:

```
SELECT
SUM(CASE WHEN transaction_type = 'credit' THEN amount END)
AS        Credit_amount, SUM(CASE WHEN transaction_type = 'debit'
THEN amount  END) AS debit_amount
FROM transactions
WHERE account_number = '824567982';
```

This query will give you the total credit and debit amount for that account, which can be useful for understanding the financial activity of that account, identifying trends in spending or income, and making decisions about loans, credit limits, and interest rates.

Credit_amount	debit_amount
500	300

Figure 8.22 – Query result

AVG()

The `AVG()` function in SQL is an aggregate function that is used to calculate the average of a set of values in a query. The `AVG()` function is commonly used with the `GROUP BY` clause to calculate the average of a set of values for each group.

The basic syntax of the `AVG()` function is as follows:

```
SELECT
AVG(column_name)
FROM table_name;
```

Please note that the AVG () function ignores NULL values when calculating the average, and the column used should be numeric.

The AVG () function is a very useful tool for data analysis and reporting, as it helps to get the central tendency of the data, which can be used to make decisions and predictions.

For example, if you wanted to calculate the average price of orders, you would use the AVG () function as follows:

```
SELECT
AVG(price)
FROM orders;
```

This query would return the average price of all the orders.

You could also use a GROUP BY clause to find the average of a specific group of orders:

```
SELECT
quantity,
AVG(price)
FROM orders
GROUP BY quantity;
```

The preceding query will return the average price of each quantity of orders.

Case scenario

An interesting use case for the AVG () function in SQL would be in the retail industry. Imagine a retail store wants to track the average sale per customer in order to identify patterns and trends in customer behavior. The store could use the AVG () function to calculate the average sale amount per customer based on data from the store's sales transactions table.

customer_id	transaction_id	sale_amount	sale_date
1	1001	7000	1/1/2023
1	1002	7000	1/2/2023
1	1003	7000	1/3/2023
1	1004	3000	2/1/2023
1	1005	3000	2/2/2023
2	2001	1250	1/1/2023
2	2002	1250	1/2/2023
2	2003	1250	2/1/2023
2	2004	1250	2/2/2023
3	3001	4000	1/1/2023
3	3002	4000	1/2/2023
3	3003	700	2/1/2023
3	3004	700	2/2/2023
4	4001	3200	1/1/2023
4	4002	3200	1/2/2023
4	4003	4000	2/1/2023
4	4004	4000	2/2/2023

Figure 8.23 – Sales transactions table

Here's an example of how this could be done:

```
SELECT
customer_id,
AVG(sale_amount) as avg_sales
FROM sales_transactions
GROUP BY customer_id;
```

This query would group the sales transactions by customer_id and calculate the average sale amount per customer. The store could then analyze this data to identify patterns in customer behavior, such as which customers tend to make larger or smaller purchases.

customer_id	avg_sales
1	5400
2	1250
3	2350
4	3600

Figure 8.24 – Query result

For example, the store could use the AVG() function to find the average sale amount per customer for each month of the year and analyze whether there is any variation in customer behavior across different seasons:

```
SELECT
    customer_id,
    DATE_FORMAT(sale_date, '%Y-%m') as month,
    AVG(sale_amount) as avg_sales
FROM sales_transactions
GROUP BY customer_id, month
ORDER BY month;
```

customer_id	month	avg_sales
1	2023-01	7000
1	2023-02	3000
2	2023-01	1250
2	2023-02	1250
3	2023-01	4000
3	2023-02	700
4	2023-01	3200
4	2023-02	4000

Figure 8.25 – Query result

An interesting use case for the AVG() function in SQL in a real-world scenario would be analyzing customer ratings for different products. You may have a table named product_ratings with the following structure:

product_id	customer_id	rating
1	1001	4.5
1	1002	3.7
2	1001	4.8
2	1003	4.2
3	1002	3.9
3	1003	4.5

Figure 8.26 – Product ratings table

In this scenario, you can use the AVG () function to calculate the average rating for each product. The AVG () function will help you understand the overall satisfaction level of customers for each product.

Here's an example SQL query that uses the AVG () function to calculate the average rating for each product:

```
SELECT product_id, AVG(rating) AS average_rating
FROM product_ratings
GROUP BY product_id;
```

The query will return the following result:

product_id	average_rating
1	4.1
2	4.5
3	4.2

Figure 8.27 – Query result

In this example, the query groups the ratings by product_id and calculates the average rating for each product using the AVG () function. The result provides insights into the average satisfaction level of customers for each product. You can use this information to identify the most highly rated products and make informed business decisions, such as promoting popular products or improving the quality of lower-rated products.

Several **Key Performance Indicators (KPIs)** can be generated to gain insights into customer satisfaction and product performance. Here are some KPIs that can be derived from the query:

- **Average Rating per Product**: This KPI represents the overall satisfaction level of customers for each product. It provides an indication of how well received a product is among customers.

- **Top-Rated Products**: By identifying the products with the highest average rating, you can determine the best-performing products in terms of customer satisfaction. This KPI helps in identifying top-performing products, which can be promoted or used as benchmarks for other products.

- **Lowest-Rated Products**: Conversely, by identifying the products with the lowest average rating, you can identify areas of improvement and potential product issues that need to be addressed. This KPI can help in understanding customer pain points and guiding product enhancements.

- **Product Rating Distribution**: By analyzing the distribution of ratings for each product, you can gain insights into the spread of customer opinions. This KPI helps in understanding whether the ratings are skewed toward either extremely positive or negative values or whether there is a balanced distribution of ratings.

- **Overall Customer Satisfaction**: By aggregating the average ratings across all products, you can calculate the overall customer satisfaction level for the entire product range. This KPI provides a high-level view of customer sentiment toward the products as a whole.

MIN() and MAX()

The `MIN()` and `MAX()` functions in SQL are aggregate functions that return the minimum and maximum values, respectively, from a column in a table.

The `MIN()` function returns the smallest value in a column, while the `MAX()` function returns the largest value in a column. Both functions can be used with any data type, including text, numbers, and dates.

The basic syntax for using the `MIN()` function is as follows:

```
SELECT MIN(column_name) FROM table_name;
```

The basic syntax for using the `MAX()` function is as follows:

```
SELECT MAX(column_name) FROM table_name;
```

For example, say we have a table named `orders` with a column named `total_price`:

order_id	total_price	order_date
1	100	1/1/2020
2	150	1/2/2020
3	200	1/3/2020
4	75	1/4/2020
5	125	1/5/2020

Figure 8.28 – orders table

To get the smallest value in the `total_price` column, we can use the following code:

```
SELECT MIN(total_price) FROM orders;
```

This will return the smallest value in the `total_price` column. The result of the query would be 75, which is the smallest value in the `total_price` column.

```
SELECT MAX(total_price) FROM orders;
```

The preceding code will return the highest value in the `total_price` column. The result of the query would be 200, which is the highest value in the `total_price` column.

You can also use both the `MIN()` and `MAX()` functions with a `WHERE` clause to filter the data and retrieve the minimum and maximum values from a specific subset of rows:

```
SELECT MIN(total_price) FROM orders WHERE order_date >= '2020-01-03';
```

The result of this query would be 125, which is the smallest value in the `total_price` column of orders placed on or after January 3, 2020.

```
SELECT MAX(total_price) FROM orders WHERE order_date >= '2020-01-03';
```

The result of the preceding query would be 200, which is the highest value in the `total_price` column of orders placed on or after January 3, 2020.

Case scenario

An interesting case scenario for using the `MIN()` and `MAX()` functions in SQL in a real-world scenario would be analyzing the performance of a company's sales team. A company could use the `MIN()` and `MAX()` functions to determine the lowest- and highest-performing salesperson based on the number of sales they have generated. The company could then use this information to provide additional support to the lowest-performing salesperson or recognize the highest-performing salesperson.

salesperson_name	date	sales_amount
John	1/1/2020	1000
John	1/2/2020	1200
Jane	1/1/2020	800
Jane	1/3/2020	900
Bob	1/1/2020	1300
Bob	1/2/2020	1400

Figure 8.29 – Sales table

A query to find this information could look like this:

```
SELECT
MIN(sales_amount) as min_sales,
MAX(sales_amount) as max_sales,
salesperson_name
FROM sales
GROUP BY salesperson_name
```

This query would return the minimum and maximum sales amount for each salesperson, along with their name, allowing the company to easily identify the lowest- and highest-performing salespeople.

min_sales	max_sales	salesperson_name
800	1400	Bob
1000	1200	John
800	900	Jane

Figure 8.30 – Query result

Case scenario 2

A very frequent use case for the MIN () and MAX () functions is to find the start and end date of a range. Let's look at this with an example.

Let's consider a scenario where you are working for a logistics company, and you have a table named shipment_schedule that tracks the scheduled shipment dates for different products. The table has the following structure:

product_id	shipment_date
1	6/1/2023
1	6/5/2023
1	6/10/2023
2	6/2/2023
2	6/8/2023

Figure 8.31 – Shipment schedule table

In this scenario, you can use the MIN () and MAX () functions to identify the start and end dates of the shipment schedule for each product. This information is helpful for determining the duration of the shipment schedule and planning logistics operations.

Here's an example SQL query that uses the `MIN()` and `MAX()` functions to identify the start and end dates for each product:

```
SELECT product_id, MIN(shipment_date) AS start_date, MAX(shipment_
date) AS end_date
FROM shipment_schedule
GROUP BY product_id;
```

The query will return the following result:

product_id	start_date	end_date
1	6/1/2023	6/10/2023
2	6/2/2023	6/8/2023

Figure 8.32 – Query result

In this example, the query groups the shipment dates by `product_id` and uses the `MIN()` function to calculate the earliest (start) date and the `MAX()` function to calculate the latest (end) date for each product. The result provides insights into the start and end dates of the shipment schedule for each product.

You can use this information to plan resources, allocate transportation, and optimize logistics operations based on the duration of the shipment schedule for different products.

In the transportation industry, a company may use the `MIN()` and `MAX()` functions to determine the earliest and latest departure times for a specific route. This information can be used to optimize schedules and improve efficiency. A query using the `MIN()` function could be used to find the earliest departure time for a particular route, while a query using the `MAX()` function could be used to find the latest departure time. This information can be used to adjust departure and arrival times to minimize delays and reduce travel time.

route_id	departure_time	arrival_time
ROUTE_123	7:00	9:30
ROUTE_123	8:00	10:30
ROUTE_123	9:00	11:30
ROUTE_456	6:00	8:30
ROUTE_456	7:00	9:30

Figure 8.33 – Transportation_schedule table

The following query will return the earliest departure time and the latest arrival time for route ROUTE_123 from the `transportation_schedule` table.

```
SELECT
MIN(departure_time) as ear_dep_time,
MAX(arrival_time) as lat_dep_time
FROM transportation_schedule
WHERE route_id = 'ROUTE_123'
```

This can be useful for determining the first and last trips of the day for a particular route, which can be used for scheduling and planning purposes.

We can get the following insights from this query:

- The minimum and maximum delivery times for a given route, which can help identify inefficiencies in the transportation process

- Additionally, the query can be used to identify the fastest and slowest drivers, which can help with scheduling and training

- Other important KPIs that can be derived from the preceding query include the number of late deliveries, the number of on-time deliveries, and the average delivery time for a given route

This information can be used to measure the performance of the transportation department and identify areas for improvement.

COUNT(DISTINCT)

The COUNT(DISTINCT) function in SQL is used to count the number of unique values in a specified column of a table. It is important because it allows you to determine the number of unique values in a column rather than just the number of total rows.

For example, consider a table named `customers` with a column named `customer_id`. If you want to find out how many unique customers the company has, you can use the following query:

```
SELECT COUNT(DISTINCT customer_id) FROM customers;
```

This query would return the number of unique `customer_id` values in the `customers` table rather than just the total number of rows. This can be useful when you want to determine the number of unique values in a column, as opposed to just the total number of rows.

In real-world scenarios, COUNT(DISTINCT) is important in many fields, such as e-commerce, finance, and marketing, where you may want to determine the number of unique customers, products, or transactions, for example. The COUNT(DISTINCT) function helps you obtain this information in a fast and efficient manner, which can be crucial for making informed business decisions.

Case scenario

Let's say we have a table named `user_posts` with `user_id`, `post_id`, and `post_text` columns. On a social media platform, we may want to know the number of unique users who have posted on the platform. We can use the `COUNT(DISTINCT)` function to achieve this:

user_id	post_id	date
1	1001	1/1/2021
1	1002	1/2/2021
2	1003	1/3/2021
3	1004	1/4/2021
3	1005	1/5/2021

Figure 8.34 – user_posts table

```
SELECT COUNT(DISTINCT user_id) as unique_users
FROM user_posts;
```

This query will return the number of unique user IDs in the `user_posts` table, allowing us to see how many distinct users have posted on the platform.

unique_users
3

Figure 8.35 – Query result

In this example, it's important to use `COUNT(DISTINCT)` instead of just `COUNT` because multiple posts could be made by the same user. If we just used `COUNT`, we would get the total number of posts, not the number of unique users. By using `COUNT(DISTINCT)`, we ensure that we only count each user once, regardless of the number of posts they have made.

> **Note**
>
> The `COUNT(DISTINCT column)` function, in more complicated queries, can have side effects and potentially impact the results by making other columns distinct as well. It's important to understand how the `COUNT(DISTINCT)` function works and its potential implications.

The `COUNT(DISTINCT column)` function counts the number of distinct values in the specified column. However, when used in a query with other columns, it affects the result set by considering the distinct values across all columns in the query. This means that if there are multiple columns in the `SELECT` statement, the `COUNT(DISTINCT column)` function could inadvertently make the entire row distinct, not just the specified column.

Consider the following example:

```
SELECT COUNT(DISTINCT customer_id) AS distinct_customers, order_date,
SUM(total_amount) AS total_revenue FROM orders GROUP BY order_date;
```

In this query, we want to count the distinct number of customers per order date while also retrieving the order date and total revenue. However, the use of `COUNT(DISTINCT customer_id)` affects the entire row, potentially making each row distinct based on the combination of `customer_id`, `order_date`, and `total_revenue`. This may not be the intended result, as we only want the distinct count for the `customer_id` column.

To avoid such side effects, it's essential to carefully structure your query and consider the impact of using `COUNT(DISTINCT)` when multiple columns are involved. You may need to refactor the query or use subqueries to achieve the desired results without unintended consequences on other columns.

In the retail industry, the `COUNT(DISTINCT)` function can be used to determine the number of unique customers who have made a purchase in a certain time period. This can help a company understand the number of new customers they have gained and how many repeat customers they have.

customer_id	purchase_date
1	12/1/2021
2	1/5/2022
3	1/15/2022
1	2/1/2022
2	3/5/2022
4	3/10/2022

Figure 8.36 – Orders table

The query would look something like this:

```
SELECT
COUNT(DISTINCT customer_id)
FROM orders
WHERE purchase_date >= DATE_SUB(NOW(), INTERVAL 3 MONTH)
```

This query would return the number of unique customers who have made a purchase in the last 3 months. In this case, the query would return 3, because there are 3 unique customer IDs in the last 3 months (customer IDs 1, 2, and 4).

A few key insights that can be generated from the query are the number of unique customers that have made a purchase or the number of unique products sold. This can provide insights into customer retention and the popularity of different products. The business could also use the `COUNT(DISTINCT)` function to analyze the number of unique store locations where sales are taking place, which can provide insights into the distribution and reach of the business. Additionally, the business could use COUNT(DISTINCT) to determine the number of unique transaction dates, which can provide insights into sales trends and patterns over time.

An interesting industry example of how GROUP BY can be used in SQL is in the field of healthcare. A hospital may have a table of patient visits, with columns for patient ID, visit date, diagnosis, and treatment.

To understand the most common diagnoses and treatments, the hospital might use GROUP BY to group the visits by diagnosis and treatment and then use aggregate functions such as COUNT() to calculate the number of visits for each diagnosis and treatment.

patient_id	visit_date	diagnosis	treatment
1	1/1/2020	Flu	Medicine
2	1/2/2020	Cold	Rest
3	1/3/2020	Flu	Medicine
4	1/4/2020	Cold	Medicine
5	1/5/2020	Flu	Rest

Figure 8.37 – patient_visits table

For example, the following SQL query would group visits by diagnosis and treatment and calculate the number of visits for each diagnosis and treatment:

```
SELECT
diagnosis,
treatment,
COUNT(*) as num_visits
FROM patient_visits
GROUP BY diagnosis, treatment
ORDER BY num_visits DESC;
```

This query would return a result set with one row for each unique combination of diagnosis and treatment in the patient_visits table, showing the number of visits for that group.

diagnosis	treatment	num_visits
Cold	Medicine	1
Cold	Rest	1
Flu	Medicine	2
Flu	Rest	1

Figure 8.38 – Query result

The hospital can use this information to identify the most common diagnoses and treatments and allocate resources accordingly, for example, adding more staff for the most common treatments or creating a specialized clinic for the most common diagnosis.

Case scenario – using all aggregate functions

Let's perform a coffee store performance review using SQL aggregate functions and help the store owners make informed decisions about the store's performance.

order_id	customer_id	order_date	total_amount	quantity
1	1001	1/1/2023	20.5	2
2	1002	1/2/2023	15.75	1
3	1003	1/3/2023	35.2	3
4	1001	1/4/2023	18.9	2
5	1002	1/5/2023	25.4	2
6	1004	1/6/2023	10	1

Figure 8.39 – Orders table

Let's say the store manager wants to generate KPIs such as total revenue, average quantity, maximum order amount, minimum order amount, and order count using the preceding dataset:

```
SELECT
    SUM(total_amount) AS total_revenue,
    AVG(quantity) AS average_quantity,
    MAX(total_amount) AS max_order_amount,
    MIN(total_amount) AS min_order_amount,
    COUNT(*) AS order_count
FROM
    orders;
```

When you execute this SQL statement with the provided sample dataset, it will generate the following output:

total_revenue	average_quantity	max_order_amount	min_order_amount	order_count
125.85	1.833333	35.2	10	6

Figure 8.40 – Results table

A single SQL statement combines multiple aggregate functions (SUM, AVG, MAX, MIN, and COUNT) to calculate and retrieve all the different KPIs simultaneously, which can help the store manager understand the overall performance, as explained here:

- **Total Revenue**: The `total_revenue` obtained from the SQL query gives the overall revenue generated by the coffee store. It provides an essential measure of the store's financial performance.

- **Average Quantity**: The `average_quantity` derived from the SQL query gives the average quantity of coffee ordered. This KPI helps in understanding the average demand for coffee products.

- **Maximum Order Amount**: The `max_order_amount` obtained from the SQL query represents the highest order amount placed by a customer. This KPI identifies the largest individual purchase made in terms of the total amount.

- **Minimum Order Amount**: The `min_order_amount` derived from the SQL query represents the lowest order amount placed by a customer. This KPI highlights the smallest individual purchase made in terms of total amount.

- **Order Count**: The `order_count` obtained from the SQL query provides the total number of orders placed. This KPI reflects the overall volume of sales and indicates customer activity.

By monitoring these KPIs over time and comparing them against targets or historical data, the coffee store manager can evaluate its progress and make informed decisions to optimize operations and improve profitability.

Summary

In this chapter, we learned how aggregate functions are used to perform calculations on a set of values and return a single value. The different aggregate functions include the following:

- `COUNT()`: Returns the number of rows in a specified column
- `SUM()`: Returns the sum of the values in a specified column
- `AVG()`: Returns the average value of the values in a specified column
- `MIN()`: Returns the minimum value in a specified column
- `MAX()`: Returns the maximum value in a specified column

These functions are often used in conjunction with a GROUP BY clause, which groups the rows in a table based on one or more specified columns. By applying aggregate functions to grouped data, you can gain insights into the characteristics of the data and make informed decisions.

For example, you can use COUNT() and AVG() to find out the number of products sold and the average price of products in a store, or you can use SUM() to calculate the total sales of a store.

It is important to note that aggregate functions ignore null values.

In summary, aggregate functions in SQL allow you to perform calculations on a set of values and return a single value, giving you insights into the characteristics of the data and helping you make informed decisions. They are often used in conjunction with a GROUP BY clause to analyze grouped data.

In the next chapter, we will learn about window functions, which are useful for performing calculations across a set of rows that are related to the current row in the result set without the need for self-joins or subqueries. They allow for advanced data analysis such as ranking, running totals, and cumulative distribution.

9

SQL Window Functions

SQL window functions aid in data wrangling by providing a way to perform complex calculations and manipulations on a set of rows related to the current row in the result set without the need for self-joins or multiple subqueries. This makes the process of data wrangling more efficient, as well as easier to understand and maintain. Window functions are comparable to the types of calculation that can be done with an aggregate function. However, we need to keep in mind that window functions can be challenging at times due to their complex syntax, so we need a deep understanding of their underlying concepts to use them properly. They require careful consideration of the window frame and the partitioning clauses, as well as the ordering of rows. Additionally, efficiently optimizing window functions can be difficult, especially with large datasets, as they often involve performing calculations over multiple rows simultaneously.

We will be covering the following main topics:

- Aggregate functions
- Window functions versus aggregate functions
- RANK, DENSE_RANK, and ROW_NUMBER
- LAG and LEAD
- SUM, AVG, MIN, and MAX
- Partitioning
- Ordering

The importance of SQL window functions

SQL window functions are important for several reasons.

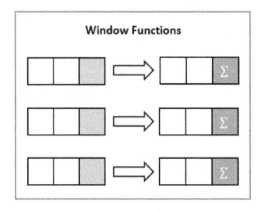

Figure 9.1 – SQL window functions

Let's explore some of them:

- **Ease of use**: Window functions simplify complex calculations by providing a convenient way to analyze a group of rows connected to the current row without requiring self-joins or multiple subqueries. This makes it easier for analysts and data scientists to perform data-wrangling tasks.

- **Increased efficiency**: Window functions enable you to perform calculations within a single query, reducing the need for multiple subqueries or self-joins. This results in improved query performance and reduced complexity.

- **Improved readability**: Window functions allow you to express complex calculations in a single query, making it easier to understand the logic and intent of the calculation. This improved readability makes it easier to maintain and modify the calculation in the future.

- **Improved data analysis**: Window Functions enable you to perform calculations such as running totals, moving averages, and rankings, which are commonly used in data analysis. This makes it possible to perform complex data analysis tasks within a single query without the need for additional code or algorithms.

Overall, SQL window functions play a crucial role in data wrangling and analysis, making it possible to perform complex calculations and manipulations on your data in an efficient, readable, and maintainable manner.

SQL aggregate functions

We've covered aggregate functions in the previous chapter, but here's a quick refresher before we compare them with window functions. SQL aggregate functions are functions that perform calculations on a set of values and return an aggregated result after summarization. These functions are used to summarize data within a specified window of rows related to the current row in the result set.

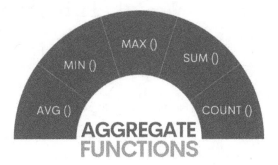

Figure 9.2 – SQL aggregate functions

Some of the most commonly used SQL aggregate functions are as follows:

- SUM: Returns the sum of values for a specified column
- AVG: Returns the average of values for a specified column
- MIN: Returns the minimum value for a specified column
- MAX: Returns the maximum value for a specified column
- COUNT: Returns the number of rows in a specified window

These aggregate functions can be used in conjunction with a window specification to perform complex calculations and manipulations on your data without the need for self-joins or multiple subqueries. For example, you could use the SUM function with a window specification to calculate the running total of sales for a particular product. Or you could use the AVG function with a window specification to compute the moving average for the temperature in a particular location over a specified period of time. Overall, SQL aggregate functions are powerful tools for data wrangling and analysis, enabling you to perform complex calculations and manipulations on your data within a single query.

SQL window functions versus aggregate functions

SQL window functions and aggregate functions are both used to perform calculations on a set of rows within a table, but they operate in different ways and are used for different purposes.

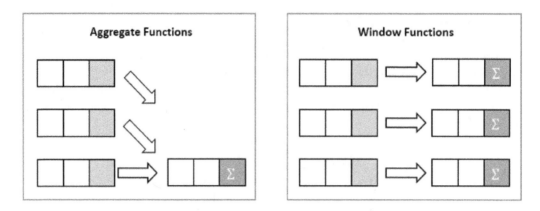

Figure 9.3 – SQL window functions versus SQL aggregate functions

Let's discuss the key differences:

- **Aggregate functions** are used to calculate a single value from a group of rows. Examples of aggregate functions are COUNT, SUM, AVG, MAX, and MIN. These functions are used with the GROUP BY clause to group rows based on one or more columns and then perform the aggregate calculation on each group. The result is a single value for each group. Aggregate functions cannot be used in SELECT clauses without a GROUP BY clause.

- **Window functions**, on the other hand, are used to calculate a value based on a window of rows within a table without grouping the rows into a single value. Window functions can be used to calculate running totals, moving averages, and other types of calculations that require access to multiple rows of data at the same time. Window functions can be used in SELECT clauses without a GROUP BY clause, and they can also be used with OVER clauses to specify the window of rows to use in the calculation.

Let's see an example to understand the differences better.

Window functions versus aggregate functions – an example to illustrate the differences

Suppose you have a table called sales with the following columns: region, year, quarter, and sales_amt. You want to calculate the total sales for each quarter in each region, as well as the percentage of the total sales that each quarter represents within each region.

To calculate the total sales for each quarter in each region, you would use an aggregate function:

```
SELECT region, year, quarter, SUM(sales_amt) as total_sales
FROM sales
GROUP BY region, year, quarter;
```

This query will return a result set with the total sales for each quarter in each region.

To calculate the percentage of the total sales that each quarter represents within each region, you would use a window function:

```
SELECT region, year, quarter, sales_amt,
       sales_amt/SUM(sales_amt) OVER (PARTITION BY region, year) as
pct_total_sales
FROM sales;
```

This query will return a result set with the percentage of total sales for each quarter in each region, calculated using a window function that partitions the rows by region and year.

In summary, while both aggregate functions and window functions are used to perform calculations on a set of rows within a table, but they have different uses and operate in different ways. Aggregate functions are used to calculate a single value from a group of rows, while window functions are used to calculate a value based on a window of rows within a table.

Window functions

Window functions in SQL are used to perform calculations on a specific subset of rows within a partition or window. They provide a way to access and manipulate data related to the current row without the need for self-joins or subqueries.

The syntax for window functions is as follows:

```
<function_name> OVER (PARTITION BY <partition_column> ORDER BY <order_
column> ROWS <window_frame>)
```

- `<function_name>` represents the window function to be applied (e.g., SUM, AVG, ROW_NUMBER)
- `PARTITION BY` allows you to divide the rows into partitions based on one or more columns
- `ORDER BY` specifies the order of the rows within each partition
- `ROWS` defines the window frame, specifying the range of rows to be included in the calculation

Overall, window functions provide a powerful way to perform calculations on subsets of rows within a partition, enhancing analytical capabilities in SQL. Let's look at different window functions:

- **Ranking functions**: Ranking functions assign a rank or row number to each row within a partition based on specified criteria. Examples are `ROW_NUMBER()`, `RANK()`, and `DENSE_RANK()`, which provide unique identifiers or rankings for each row.
- **Aggregate functions**: Window versions of aggregate functions (e.g., `SUM()`, `AVG()`, and `COUNT()`) calculate aggregate values over a window rather than the entire dataset. They allow calculations such as the running total or moving average within the specified window.

- **Analytic functions**: Analytic functions provide insights and additional information about a specific row by computing values based on the values within the window. Examples are LAG(), LEAD(), and FIRST_VALUE(), which retrieve the values from preceding or subsequent rows.

- **Percentile functions**: Percentile functions (e.g., PERCENT_RANK(), NTILE()) help determine the relative position of a row within a partition based on sorted values. They can be used to identify the distribution of data and to determine percentiles.

- **Offset functions**: Offset functions (e.g., LAG(), LEAD()) retrieve values from the preceding or subsequent rows within a partition. They are useful for comparing values or calculating the difference between consecutive rows.

These different types of window functions provide flexibility in performing calculations and analysis on a subset of rows, allowing comparisons, rankings, aggregations, and other insights within the specified window or partition. Let's take a deeper dive into the most important window functions.

SUM()

The SUM() SQL function is an aggregate function that returns the sum of values for a specified column within a specified window of rows related to the current row in the result set. This function allows you to perform running total calculations, cumulative sums, and other types of calculations that involve the addition of values within a set of rows.

The syntax for using the SUM() function is as follows:

```
SUM(column_name) OVER (PARTITION BY partition_column ORDER BY order_
column [ROWS BETWEEN start_row AND end_row])
```

The PARTITION BY clause is used to specify the subset of rows to be used in the calculation, based on one or more columns. The ORDER BY clause is used to specify the order in which the rows are processed for the calculation. The ROWS BETWEEN clause is used to specify the range of rows to include in the calculation, relative to the current row.

For example, you could use the SUM() function to calculate the running total of sales for each product in a table:

```
SELECT
product_id,
sale_date,
sales,
SUM(sales) OVER (PARTITION BY product_id ORDER BY sale_date) as
running_total
FROM sales_table;
```

This query returns the running total of sales for each product, which has been calculated by adding the sales for each row to the running total of sales for the previous row.

Overall, the SUM () function is a valuable tool for data wrangling and analysis, enabling you to perform complex calculations and manipulations on your data within a single query.

Scenario

The SUM () function can be useful in various scenarios in the healthcare industry. Let's discuss one such scenario.

Suppose you have a table of patient medical records, which includes information such as patient ID, date of visit, and the cost of medical services provided. You want to calculate the cumulative cost of medical services for each patient over time to track the financial impact of medical treatment on each patient.

patient_id	visit_date	cost
1	1/1/2020	100
1	1/2/2020	200
1	1/3/2020	150
2	2/1/2020	50
2	2/2/2020	75
3	3/1/2020	250

Figure 9.4 – Patient data

You could use the SUM () function to calculate the cumulative cost of medical services for each patient, as follows:

```
SELECT
patient_id,
visit_date,
cost,
SUM(cost) OVER (PARTITION BY patient_id ORDER BY visit_date) as
cumulative_cost
FROM medical_records;
```

This query returns the cumulative cost of medical services for each patient, calculated by adding the cost of medical services for each row to the cumulative cost of medical services for the previous row. This information could be used to track the financial impact of medical treatment on each patient and to make decisions about future medical treatment based on financial considerations.

patient_id	visit_date	cost	cumulative_cost
1	1/1/2020	100	100
1	1/2/2020	200	300
1	1/3/2020	150	450
2	2/1/2020	50	50
2	2/2/2020	75	125
3	3/1/2020	250	250

Figure 9.5 - Query result

So, for patient 1, the cumulative cost is calculated as the sum of the cost of medical services from all their previous visits plus the cost of medical services from the current visit. The same is done for other patients.

Overall, the SUM () function is a useful tool for healthcare data analysis. It allows you to perform complex calculations and manipulations on your data in a simple and intuitive manner.

Let's look at another interesting example. Suppose you have a table of farm production data, which contains farm ID, crop type, date of harvest, and the yield (in pounds) of each crop data. You want to calculate the cumulative yield of each crop type for each farm over time to track the performance and productivity of each farm.

You could use the SUM () function to calculate the cumulative yield of each crop type for each farm as follows:

```
SELECT farm_id, crop_type, harvest_date, yield, SUM(yield) OVER
(PARTITION BY farm_id, crop_type ORDER BY harvest_date) as cumulative_
yield
FROM farm_production;
```

This query returns the cumulative yield of each crop type for each farm, calculated by adding the yield from each harvest to the cumulative yield from all previous harvests. This information could be used to track the performance and productivity of each farm and to make decisions about future crop types to sow based on yield considerations.

farm_id	crop_type	harvest_date	Yield	cumulative_yield
1	Wheat	5/1/2020	500	500
1	Wheat	5/15/2020	700	1200
1	Corn	6/1/2020	800	800
2	Wheat	5/10/2020	400	400
2	Corn	6/5/2020	600	600

Figure 9.6 – Farm production data

Overall, the SUM() function is a useful tool for cumulative data analysis. It allows you to perform complex calculations and manipulations on your data in a simple and intuitive manner.

COUNT()

The COUNT() function can also be used with the OVER clause and a PARTITION BY clause to perform a grouped count, that is, to count the number of rows in each group of a result set. The PARTITION BY clause is used to divide the result set into partitions, or groups, based on the values of one or more columns.

Here is the basic syntax for the COUNT() function with the OVER and PARTITION BY clauses:

```
SELECT column1, column2, COUNT(column3) OVER (PARTITION BY column4)
FROM table_name;
```

In this syntax, column1, column2, and column3 are the names of columns that you want to retrieve data from. column4 is the name of the column that you want to use to divide the result set into partitions. table_name is the name of the table that you want to retrieve data from.

order_id	customer_name	order_date
1	Seoge Ui	1/1/2022
2	Seoge Ui	1/1/2022
3	Seoge Ui	1/1/2022
4	Hary Sae	1/2/2022
5	Hary Sae	1/2/2022

Figure 9.7 – Order table

Here is an example that demonstrates the use of the COUNT() function with the OVER and PARTITION BY clauses:

```
SELECT customer_name, order_date, COUNT(order_id) OVER (PARTITION BY
customer_name)
FROM orders;
```

This query returns customer_name, order_date, and the number of orders for each customer. That is, it returns the number of orders placed by each customer. The result set is divided into partitions based on the customer_name column, and the COUNT() function is used to count the number of orders for each partition.

customer_name	order_date	count
Seoge Ui	1/1/2022	3
Hary Sae	1/2/2022	2

Figure 9.8 – Result table

As you can see, the result set is divided into partitions based on the customer_name column. The COUNT() function is used to count the number of orders for each partition and the result is displayed in the count column. The partitioning is specified by the OVER clause and defined by the PARTITION BY clause. The query returns customer_name, order_date, and the number of orders placed by each customer.

Scenario

In the social media industry, the COUNT() function in SQL can be used to analyze the engagement of users with posts. For example, consider a table called Posts with Post_ID, User_ID, Post_Content, and Timestamp columns.

Post_ID	User_ID	Post_Content	Timestamp
1	100	Hello world!	2/1/2022 12:00
2	100	Nice weather!	2/2/2022 8:00
3	200	My new pet	2/2/2022 10:00
4	200	#Throwback	2/3/2022 15:00
5	300	Check this out	2/4/2022 13:00
6	300	Happy Friday!	2/4/2022 17:00

Figure 9.9 – User engagement

To count the number of posts each user has made, you could use the following SQL query:

```
SELECT User_ID, COUNT(Post_ID) OVER(PARTITION BY User_ID) AS Post_
Count FROM Posts
```

This query creates a result set that includes two columns, User_ID and Post_Count. The Post_Count column contains the number of posts made by each user, and the PARTITION BY clause ensures that the count is calculated separately for each user.

User_ID	Post_Count
100	2
200	2
300	2

Figure 9.10 – Result table

You could then use this data to identify top contributors, monitor user activity, or identify trends in post frequency over time.

The **key performance indicator (KPI)** generated by the preceding analysis would be *Number of Posts per User*. This KPI provides insight into the level of engagement of users on the social media platform, as well as the frequency and volume of content being created by users.

Leadership can use this KPI to track the growth of the platform over time and identify patterns or trends in user behavior. They can also use it to identify top contributors, influencers, or popular topics, which can be used for marketing or advertising purposes.

Additionally, this KPI can be used to set targets or benchmarks for user engagement and content creation, and to track progress towards those goals. It can also help identify areas where user engagement may be lacking, which can be addressed through targeted campaigns or initiatives to drive user activity and content creation. Overall, the *Number of Posts per User* KPI is a valuable metric that can provide valuable insights into user engagement and help inform strategic decisions for the social media platform.

There are several other KPIs that can be generated using the `Posts` dataset. Here are a few examples:

- **Engagement Rate**: This KPI measures the percentage of users who engage with a post by liking, commenting, or sharing it. It can be calculated as *(Total Engagements / Total Views) * 100*. This KPI can identify the posts that are resonating with users and driving engagement.

- **Active Users**: This KPI measures the number of unique users who have created a post within a specific time period. It can help track user activity and engagement over time and identify trends in user behavior.

- **Response Time**: This KPI measures the time it takes for a user to respond to a post or message. It can help identify areas where customer service or support may need improvement and ensure timely responses to user inquiries or issues.

- **Reach**: This KPI measures the number of unique users who have viewed a post or message. It can help identify the overall reach and impact of content on the social media platform.

- **Conversion Rate**: This KPI measures the percentage of users who take a desired action, such as making a purchase or signing up for a service, after viewing a post or message. It can help track the effectiveness of marketing campaigns and identify areas for improvement.

These KPIs can provide valuable insights into user engagement, content effectiveness, and overall platform performance, and can be used to inform strategic decisions and drive growth on the social media platform.

One interesting example of the usage of the `COUNT()` function in government is to analyze crime data to help law enforcement agencies identify crime hotspots and allocate resources effectively.

For instance, the **Federal Bureau of Investigation (FBI)** collects crime data from local law enforcement agencies across the United States and uses SQL to analyze it. They can use the `COUNT()` function to calculate the total number of crimes committed in each state, county, or city, as well as the number of crimes committed in specific categories, such as violent crimes or property crimes.

City	Crime Type
New York	Robbery
New York	Assault
New York	Robbery
Los Angeles	Assault
Los Angeles	Burglary
Chicago	Assault
Chicago	Robbery
Chicago	Robbery
Chicago	Burglary

Figure 9.11 – Crime data

This data can be used to identify areas where crime rates are high and allocate law enforcement resources accordingly. For example, if a certain city or neighborhood has a high number of violent crimes, law enforcement agencies can assign more police officers to patrol that area.

In this example, we have a simple dataset of crime incidents in three different cities: New York, Los Angeles, and Chicago. The dataset includes the name of the city and the type of crime that was committed.

Using the COUNT() function, we can easily calculate the total number of crimes committed in each city. Here's an example SQL query:

```
SELECT City, COUNT(*) OVER (PARTITION BY City) AS TotalCrimes
FROM CrimeData;
```

The result of this query is as follows:

Row Labels	Count of Crime Type
Chicago	4
Los Angeles	2
New York	3

Figure 9.12 – Result data

In addition to using the COUNT() function, other SQL functions, such as SUM() and AVG(), can be used to calculate the total and average values of crime data for different areas and crime categories.

This information can be used to make data-driven decisions and improve public safety. Overall, the COUNT() function in SQL is a powerful tool for analyzing crime data, informing decisions that can reduce crime rates and make communities safer.

AVG()

The AVG() function in SQL is an important tool for data wrangling as it allows us to calculate the average value of a specific column over a certain window of data.

This function is particularly useful when we want to calculate moving averages or rolling averages on time-series data. By using the AVG() function with a window function such as ROWS BETWEEN, we can calculate the average value of a column over a specified range of rows.

Here's an example to illustrate how the AVG() function can be used for data wrangling. Suppose we have a dataset of daily stock prices for a particular company, and we want to calculate a 7-day moving average of the stock prices.

Date	Stock Price
1/1/2022	10
1/2/2022	12
1/3/2022	15
1/4/2022	16
1/5/2022	14
1/6/2022	13
1/7/2022	12.5
1/8/2022	13

Figure 9.13 – Stock price table

Using the AVG() function, we can calculate the 7-day moving average of the stock prices. Here's an example SQL query:

```
SELECT Date, StockPrice, AVG(StockPrice) OVER (ORDER BY Date ROWS
BETWEEN 6 PRECEDING AND CURRENT ROW) AS MovingAvg FROM StockData;
Note : ROWS BETWEEN 6 PRECEDING AND CURRENT ROW syntax defines the
window size. It includes the current row and the 6 preceding rows. In
other words, it calculates the average over the last 7 days (including
the current day).
```

Here's the result:

Date	Stock Price	Moving Avg
1/1/2022	10	10
1/2/2022	12	11
1/3/2022	15	12.33
1/4/2022	16	13.25
1/5/2022	14	13.4
1/6/2022	13	13.5
1/7/2022	12.5	13.43
1/8/2022	13	13.25

Figure 9.14 – Result data

As you can see, the AVG() function has been used to add a new column to the dataset that shows the 7-day moving average of the stock prices. This information can identify trends, which will help with making informed decisions about buying or selling the company's stock.

In addition to moving averages, the AVG() function can be used to calculate rolling averages and other types of aggregations on time-series data. This makes it an important tool for data wrangling and analysis in a variety of industries.

Scenario

One scenario where it is essential to use the AVG() SQL function is when analyzing sales data for a retail company. The AVG() function can be used to calculate the average sales for each store over a specific period, allowing the company to identify high-performing and underperforming stores.

For instance, let's say that a retail company has stores in multiple locations, and it wants to determine the average daily sales for each store for the past week. It can use the AVG() function to calculate the average sales for each store and compare the results.

Store Location	Date	Daily Sales
Store A	2/1/2023	1000
Store A	2/2/2023	1200
Store A	2/3/2023	1500
Store B	2/1/2023	800
Store B	2/2/2023	900
Store B	2/3/2023	1000

Figure 9.15 – Store data

The SQL query for this could be as follows:

```
SELECT store_location, AVG(daily_sales) OVER (PARTITION BY store_
location) AS avg_sales
FROM sales_data
WHERE date >= '2023-02-01' AND date <= '2023-02-07'
ORDER BY store_location;
```

In this example, the PARTITION BY clause is used to group the sales data by store location, and the AVG() function is used to calculate the average sales for each store. The result is a table that shows the store location and the average daily sales for each store for the past week.

Row Labels	Average of Daily Sales
Store A	1233.333
Store B	900

Figure 9.16 – Result data

This information can be used to identify stores that are performing well and stores that may need additional support to improve sales. It can also be used to set sales targets and measure performance over time. In this way, the AVG() function is an essential tool for data wrangling and analysis in the retail industry.

Here's an example of how the AVG() function can be used to calculate the average transaction value in the finance industry, specifically for a hypothetical trading volume dataset:

Trade ID	Trade Date	Symbol	Quantity	Price
1	1/1/2022	AAPL	100	150
2	1/2/2022	MSFT	50	200
3	1/3/2022	GOOG	75	175
4	1/4/2022	TSLA	200	100
5	1/5/2022	AAPL	150	175
6	1/6/2022	MSFT	100	190
7	1/7/2022	TSLA	75	250
8	1/8/2022	GOOG	50	180

Figure 9.17 – Before trading volume data

In the before dataset, we have simple trading volume data that shows the trade ID, trade date, symbol, quantity, and price for each transaction. We want to calculate the average transaction value for each symbol using the AVG() function.

To do this, we can use the following SQL query:

```
SELECT Symbol, AVG(Quantity * Price) OVER (PARTITION BY Symbol) AS
'Avg Transaction Value'
FROM trading_volume_data
```

This query calculates the product of the quantity and price for each trade, then uses the AVG() Over function to calculate the average transaction value for each symbol. The PARTITION BY clause ensures that the average is calculated separately for each symbol.

The following table shows the result after using the AVG() function:

Symbol	Avg Transaction Value
AAPL	163.33
MSFT	195
GOOG	177.5
TSLA	164.29

Figure 9.18 – After AVG transaction data

The resulting after dataset shows the symbol and the corresponding average transaction value for that symbol. We can see that AAPL has an average transaction value of $163.33, MSFT has an average transaction value of $195.00, GOOG has an average transaction value of $177.50, and TSLA has an average transaction value of $164.29.

The KPIs that can be generated using this analysis include the following:

- **Average Transaction Value by Symbol**: This KPI provides insights into the average transaction value for each symbol, which can help financial leaders to identify the most profitable symbols and make informed decisions on pricing strategies

- **Total Transaction Volume by Symbol**: Financial leaders can use this KPI to identify the symbols that generate the highest trading volume and revenue

- **Average Trade Size by Symbol**: This KPI can provide insights into the typical trade size for each symbol, which can help financial leaders to determine the level of market interest in a particular symbol

- **Market Share by Symbol**: This KPI can provide insights into the market share of each symbol and help financial leaders to identify opportunities for growth and competition in the market

ROW_NUMBER()

ROW_NUMBER() is a window function in SQL that assigns a unique sequential number to each row within a result set. It is important because it allows us to add a unique identifier to each row in a table or query result. This can be useful for a variety of purposes:

- **Pagination**: When returning large result sets, we can use the ROW_NUMBER() function to partition the data into small chunks and return a specific range of rows based on the page number and page size

- **Ranking**: We can use the ROW_NUMBER() function to assign a rank to each row based on specified criteria, such as order amount, customer ID, or product type, which helps with removing duplicates

- **Filtering**: We can use the ROW_NUMBER() function to filter out specific rows based on their position within the result set, such as the top 10 highest or lowest values

Overall, the ROW_NUMBER() function is a powerful data-wrangling tool in SQL that allows us to add sequential numbering to rows and perform various operations on them.

Let's see a basic example of using the ROW_NUMBER() function in SQL. Suppose you have a table called employees with the following data:

employee_id	employee_name	department	Salary
1	John	Sales	50000
2	Alice	Marketing	60000
3	Bob	Sales	55000
4	Sarah	Marketing	65000
5	David	HR	45000

Figure 9.19 – employees table

You can use the ROW_NUMBER() function to generate a sequential number for each row in the result set. Here's an example query that uses the ROW_NUMBER() function to rank employees by salary within their respective departments:

```
SELECT employee_id, employee_name, department, salary, ROW_NUMBER()
OVER (PARTITION BY department ORDER BY salary DESC) AS rank
FROM employees;
```

This query will return a result set with the following columns:

employee_id	employee_name	department	salary	rank
1	John	Sales	50000	2
3	Bob	Sales	55000	1
2	Alice	Marketing	60000	2
4	Sarah	Marketing	65000	1
5	David	HR	45000	1

Figure 9.20 – Result data

The ROW_NUMBER() function has been used to generate a sequential number (or rank) for each row in the result set, starting from 1. The PARTITION BY clause has been used to group the rows by department, and the ORDER BY clause has been used to order the rows by salary within each department. Therefore, the rank column shows the rank of each employee within their respective department based on their salary.

The ROW_NUMBER() function is a useful tool for generating sequential numbers or ranks within a result set. It can be used to perform various analytical tasks, such as ranking, pagination, and grouping.

Scenario

Suppose we have a dataset of trucking routes, with the following columns: route_id, origin, destination, distance, and time_to_complete.

Shipment ID	Carrier	Origin	Destination	Shipment Date	Arrival Date
1	FedEx	Dallas	New York	1/1/2022	1/5/2022
2	UPS	Los Angeles	Miami	1/3/2022	1/7/2022
3	FedEx	Chicago	Miami	1/5/2022	1/10/2022
4	UPS	Boston	Dallas	1/7/2022	1/11/2022
5	FedEx	Miami	Los Angeles	1/9/2022	1/12/2022

Figure 9.21 – Shipment data

We want to add a sequential order to each route based on the time it takes to complete the route. We can use the ROW() NUMBER function to accomplish this. Here's the SQL query we can use:

```
SELECT
    Shipment_ID,
    Carrier,
    Origin,
```

```
    Destination,
    Shipment_Date,
    Arrival_Date,
    ROW_NUMBER() OVER (ORDER BY Arrival_Date - Shipment_Date) AS
  'Delivery Rank'
  FROM shipment_data
```

This query calculates the difference between the arrival date and the shipment date for each shipment and then uses the ROW_NUMBER() function to rank the shipments based on this difference. The ORDER BY clause ensures that the shipments are ranked in ascending order by the difference between the arrival date and the shipment date.

Shipment ID	Carrier	Origin	Destination	Shipment Date	Arrival Date	Delivery Rank
1	FedEx	Dallas	New York	1/1/2022	1/5/2022	1
2	UPS	Los Angeles	Miami	1/3/2022	1/7/2022	2
3	FedEx	Chicago	Miami	1/5/2022	1/10/2022	3
4	UPS	Boston	Dallas	1/7/2022	1/11/2022	4
5	FedEx	Miami	Los Angeles	1/9/2022	1/12/2022	5

Figure 9.22 – Result data

The resulting after dataset shows the shipment ID, carrier, origin, destination, shipment date, arrival date, and delivery rank for each shipment. The Delivery Rank column shows the rank of each shipment based on the difference between the arrival date and the shipment date. In this example, we can see that the first shipment by FedEx had the best on-time delivery performance with a delivery rank of 1, and the last shipment by FedEx had the worst on-time delivery performance with a delivery rank of 5.

Transportation companies can use this information to measure the performance of their carriers, identify trends and patterns, and make data-driven decisions to optimize their operations and improve on-time delivery performance.

Let's explore an interesting example of how this function can be used in the car sales and sales agent domains.

A car-selling company can use the ROW() function in SQL to rank their salespeople based on the number of cars they sold. Suppose the company has a table named sales that contains data about the cars sold, including the name of the salesperson who made each sale:

sale_id	car_id	salesperson_name	sale_date
1	101	John Smith	1/1/2022
2	102	Jane Doe	1/1/2022
3	103	John Smith	1/2/2022
4	104	John Smith	1/3/2022
5	105	Jane Doe	1/4/2022
6	106	Jack Johnson	1/5/2022
7	107	John Smith	1/6/2022

Figure 9.23 – Sales data

The company wants to rank its salespeople based on the number of cars they have sold. They can use the ROW() function in SQL to generate a ranking for each salesperson. Here's the SQL code to achieve this:

```
SELECT
  salesperson_name,
  COUNT(*) AS cars_sold,
  ROW_NUMBER() OVER (ORDER BY COUNT(*) DESC) AS sales_rank
FROM sales
GROUP BY salesperson_name;
```

This query groups the sales table by the salesperson_name column, and then uses the COUNT(*) function to count the number of cars sold by each salesperson. The ROW_NUMBER() function is used to assign a rank to each salesperson based on the number of cars they have sold. The ORDER BY clause in the ROW_NUMBER() function sorts the results in descending order by the number of cars sold.

The resulting table looks like this:

salesperson_name	cars_sold	sales_rank
John Smith	3	1
Jane Doe	2	2
Jack Johnson	1	3

Figure 9.24 – Result data

As we can see, John Smith has sold the most cars and is ranked number one, followed by Jane Doe and Jack Johnson.

The car-selling company can use this ranking to incentivize their sales team to sell more cars, as well as to reward the top performers. They can also use this data to identify areas where salespeople may need additional training or support.

Using ROW_NUMBER() to eliminate duplicates

In SQL, the ROW_NUMBER () function can be used to assign a unique sequential number to each row in a dataset in a specified order. By utilizing this function, you can eliminate duplicates from a raw dataset. Let's look at an example.

Let's say we have a table called Employees with the following columns: EmployeeID, FirstName, LastName, and Salary. The goal is to eliminate duplicate records based on the FirstName and LastName columns.

EmployeeID	FirstName	LastName	Salary
1	John	Doe	50000
2	Jane	Smith	60000
3	John	Doe	55000
4	Alex	Johnson	45000
5	Jane	Smith	65000

Figure 9.25 – Employees table

Let's see how we can use the ROW_NUMBER () function to achieve this:

```
SELECT EmployeeID, FirstName, LastName, Salary
FROM (
    SELECT EmployeeID, FirstName, LastName, Salary,
        ROW_NUMBER() OVER (PARTITION BY FirstName, LastName ORDER BY
EmployeeID) AS RowNum
    FROM Employees
) AS Subquery
WHERE RowNum = 1;
```

Here's how it works step by step:

1. Start with the inner query that assigns row numbers to each row based on FirstName and LastName:

    ```
    SELECT EmployeeID, FirstName, LastName, Salary,
        ROW_NUMBER() OVER (PARTITION BY FirstName, LastName ORDER BY
    EmployeeID) AS RowNum
    FROM Employees;
    ```

The result of this query is as follows:

EmployeeID	FirstName	LastName	Salary	RowNum
1	John	Doe	50000	1
3	John	Doe	55000	2
4	Alex	Johnson	45000	1
2	Jane	Smith	60000	1
5	Jane	Smith	65000	2

Figure 9.26 – Employees table

2. Now, in the outer query, we select only the rows where **RowNum** is equal to 1:

```
SELECT EmployeeID, FirstName, LastName, Salary
FROM (
    SELECT EmployeeID, FirstName, LastName, Salary,
        ROW_NUMBER() OVER (PARTITION BY FirstName, LastName
ORDER BY EmployeeID) AS RowNum
    FROM Employees
) AS Subquery
WHERE RowNum = 1;
```

The result of this query would be the final dataset with duplicates eliminated:

EmployeeID	FirstName	LastName	Salary
1	John	Doe	50000
4	Alex	Johnson	45000
2	Jane	Smith	60000

Figure 9.27 – Employee table without duplicates

As you can see, only the first occurrence of each unique combination of FirstName and LastName is retained, and any duplicates are removed. Thus, by using the ROW_NUMBER() function in this manner, you can effectively remove duplicates based on specific criteria in your SQL queries.

RANK() and DENSE_RANK()

Both the RANK() and DENSE_RANK() functions in SQL are used to assign a rank to each row within a result set based on the values in one or more columns. The main difference between the two functions is that RANK() leaves gaps in the ranking sequence when there are ties, whereas DENSE_RANK() does not.

For example, if two rows have the same ranking value, RANK() would assign them a tied rank and leave a gap for the next rank number, while DENSE_RANK() would assign the same rank value to both rows and not leave a gap.

Name	Score
Alice	95
Bob	85
Cindy	95
Dave	80
Emily	90

Figure 9.28 – Student data

If we want to rank the students by their scores, we can use the RANK() and DENSE_RANK() functions as follows:

```
SELECT
    name,
    score,
    RANK() OVER (ORDER BY score DESC) AS rank,
    DENSE_RANK() OVER (ORDER BY score DESC) AS denserank
FROM
    students;
```

This query will produce the following result set:

Name	Score	Rank	Denserank
Alice	95	1	1
Cindy	95	1	1
Emily	90	3	2
Bob	85	4	3
Dave	80	5	4

Figure 9.29 – Result data

As you can see, both RANK() and DENSE_RANK() assign the highest rank value of 1 to Alice and Cindy, who have the highest score of 95. However, RANK() assigns the next rank value of 3 to Emily, since it leaves a gap after the tied Alice and Cindy. In contrast, DENSE_RANK() assigns the next rank value of 2 to Emily, since it does not leave any gaps in the ranking sequence.

Similarly, RANK() assigns the next rank value of 4 to Bob, and the final rank value of 5 to Dave, while DENSE_RANK() assigns the next rank value of 3 to Bob, and the final rank value of 4 to Dave.

RANK() versus DENSE_RANK()

When choosing between RANK() and DENSE_RANK() for data wrangling, the decision depends on how you want to handle ties in your ranking.

RANK() assigns a unique rank to each row in the result set, with no gaps in the ranking sequence. If there are ties in the data, RANK() will skip the next rank(s) to account for the tie(s). For example, if there are two movies tied for first place, the next movie will be ranked third, rather than second.

DENSE_RANK() is similar to RANK(), but it does not skip any ranks when there are ties.

Here are a few scenarios where you might prefer one function over the other:

- If you're interested in finding the top *n* records (e.g. top 5, top 10, etc.), you might want to use RANK(), as it will give you a list of unique records. If you use DENSE_RANK(), you may end up with more records than you intended if there are ties.

- If you're interested in finding records with the highest or lowest value in a certain column and you want to give all tied records the same ranking, DENSE_RANK() might be more appropriate.

- If you're interested in finding the percentile rank of each record within a group, DENSE_RANK() might be more appropriate, as it will give all records in the same percentile the same ranking.

Ultimately, the choice between RANK() and DENSE_RANK() depends on the specifics of your use case and how you want to handle ties in the ranking.

Scenario

One interesting use of the RANK() and DENSE_RANK() functions is in analyzing the performance of movies in the film industry.

For example, a data scientist used these functions to rank and visualize the top-grossing movies of all time:

Movie	Worldwide Box Office Earnings
Avengers: Endgame	$2,798,000,000
Avatar	$2,789,700,000
Titanic	$2,187,500,000
Star Wars: The Force Awakens	$2,068,200,000
Avengers: Infinity War	$2,048,000,000
Jurassic World	$1,670,400,000
The Lion King	$1,656,000,000
The Avengers	$1,519,600,000
Furious 7	$1,516,000,000
Frozen 2	$1,450,000,000
Avengers: Age of Ultron	$1,402,800,000

Figure 9.30 – Movie data

Here is the SQL code that can be used to achieve ranking using the RANK() and DENSE_RANK() functions:

```
SELECT Movie, Worldwide_Box_Office_Earnings,
       RANK() OVER (ORDER BY Worldwide_Box_Office_Earnings DESC) as
Rank,
       DENSE_RANK() OVER (ORDER BY Worldwide_Box_Office_Earnings DESC)
as Dense_Rank
FROM Movies
```

Movie	Worldwide Box Office Earnings	Rank	Dense Rank
Avengers: Endgame	$2,798,000,000	1	1
Avatar	$2,798,000,000	1	1
Titanic	$2,187,500,000	3	2
Star Wars: The Force Awakens	$2,068,200,000	3	2
Avengers: Infinity War	$2,048,000,000	5	3
Jurassic World	$1,670,400,000	6	4
The Lion King	$1,656,000,000	7	5
The Avengers	$1,519,600,000	8	6
Furious 7	$1,516,000,000	9	7
Frozen 2	$1,450,000,000	10	8
Avengers: Age of Ultron	$1,402,800,000	11	9

Figure 9.31 – Worldwide box office earnings

As you can see, the RANK() function has assigned a unique rank value to each movie based on its worldwide box office earnings. The highest-grossing movie, **Avengers: Endgame**, has a rank value of 1, while the lowest-grossing movie in the dataset, **Avengers: Age of Ultron**, has a rank value of 11.

The DENSE_RANK() function, on the other hand, has assigned the same rank value to movies with tied earnings, resulting in no gaps in the ranking sequence. For example, **Avatar** and **Avengers: Endgame** have the same earnings and are tied for the top rank, with a dense rank value of 1. Similarly, **Titanic** and **Star Wars: The Force Awakens** are tied for the third rank with a dense rank value of 2.

Here is an example of how you can visualize the rankings of movies using a scatter plot.

Figure 9.32 – Scatter plot

Each movie is represented by a point on the scatter plot. Here's an example of what the scatter plot could look like for the top 10 movies in our dataset, with rankings based on worldwide box office earnings. As you can see, the scatter plot provides a visual representation of the relationship between the rankings of movies and their worldwide box office earnings. We can see that there is a general trend of higher earnings for movies with higher rankings, although there are some outliers, such as **Jurassic World** (rank 4), which has a relatively high ranking, but lower earnings compared to some other movies.

Lead() and Lag()

The Lead and Lag functions are analytical functions in SQL that allow you to access data from the next or previous row within a result set, respectively.

The Lead function returns the value of a column in the next row within the partition, while the Lag function returns the value of a column in the previous row within the partition.

Both functions take three arguments:

- The column to evaluate
- The offset (the number of rows to move forward or backward)
- The default value to return if there is no next or previous row

Let's use an example to see how to use the Lead and Lag functions in SQL.

Suppose we have a table called Sales with the following data:

Month	Sales
Jan	100
Feb	150
Mar	200
Apr	175
May	225

Figure 9.33 – Sales data

To get the sales figures for the previous and next month for each row, we can use the Lag and Lead functions as follows:

```
SELECT Month, Sales,
       Lag(Sales, 1, 0) OVER (ORDER BY Month) AS PrevMonthSales,
       Lead(Sales, 1, 0) OVER (ORDER BY Month) AS NextMonthSales
FROM Sales;
```

This will return the following result:

Month	Sales	PrevMonthSales	NextMonthSales
Jan	100	0	150
Feb	150	100	200
Mar	200	150	175
Apr	175	200	225
May	225	175	0

Figure 9.34 – Result data

In this example, we used the Lag function to get the previous month's sales figures and the Lead function to get the next month's sales figures. The third argument in each function is the default value to return if there is no previous or next row.

Overall, the Lead and Lag functions are useful tools for analyzing time-series data and calculating trends or changes in values over time.

Scenario

Here's an example dataset and some use cases that show how to use the Lead and Lag functions to wrangle this data.

Suppose we have a dataset that tracks the number of orders and the total revenue generated by a company over a period of five days:

Date	Orders	Revenue
2/1/2022	10	1000
2/2/2022	15	1500
2/3/2022	20	2000
2/4/2022	18	1800
2/5/2022	25	2500

Figure 9.35 – Orders data

We can use the Lead and Lag functions to calculate various metrics for this dataset, such as the change in the number of orders and revenue each day.

To calculate the daily change in orders, we can use the Lag function to retrieve the number of orders from the previous day and subtract it from the current day's value:

```
SELECT Date, Orders,
       Orders - Lag(Orders, 1, 0) OVER (ORDER BY Date) AS
```

```
DailyOrderChange
FROM OrdersData;
```

This will return the following result:

Date	Orders	DailyOrderChange
2/1/2022	10	0
2/2/2022	15	5
2/3/2022	20	5
2/4/2022	18	-2
2/5/2022	25	7

Figure 9.36 – Result data

To calculate the percentage change in revenue each day, we can use the Lead function to retrieve the revenue from the next day and calculate the percentage change from the current day's value:

```
SELECT Date, Revenue,
(Lead(Revenue, 1, 0) OVER (ORDER BY Date) - Revenue) / Revenue * 100
AS DailyRevenueChange
FROM OrdersData;
```

This will return the following result:

Date	Revenue	DailyRevenueChange
2/1/2022	1000	50
2/2/2022	1500	33.3
2/3/2022	2000	-11.1
2/4/2022	1800	38.9
2/5/2022	2500	0

Figure 9.37 – Result data

In this example, we used the Lag and Lead functions to calculate daily changes in orders and revenue, which can be useful for identifying trends and patterns in the data.

Here's an example of how to use the Lead and Lag SQL functions to analyze user behavior on a website using sample data. Suppose you have a table called user_sessions with the following columns:

- session_id: A unique identifier for each user session
- page_url: The URL of the page visited by the user

- `time_spent`: The time (in seconds) spent by the user on the page
- `clicks`: The number of clicks made by the user on the page

Here's what the table might look like:

session_id	page_url	time_spent	clicks
1	`/homepage.html`	30	2
1	`/products.html`	60	4
1	`/checkout.html`	15	1
2	`/homepage.html`	45	3
2	`/about.html`	20	1
3	`/homepage.html`	90	5
3	`/products.html`	30	2
3	`/checkout.html`	45	3

Figure 9.38 – User session table

Using the `Lead` and `Lag` functions, we can calculate the bounce rate, a metric that measures the percentage of users who leave the website after visiting a particular page:

```
SELECT page_url,
       1 - (COUNT(*) FILTER (WHERE Lag(page_url, 1, '') OVER
(PARTITION BY session_id ORDER BY time_spent) = page_url) / COUNT(*)
OVER (PARTITION BY session_id)) AS bounce_rate
FROM user_sessions
GROUP BY page_url
ORDER BY bounce_rate DESC;
```

This query calculates the bounce rate for each page visited by users. The `Lag` function is used to retrieve the previous page visited by each user during their session, and the `COUNT` and `FILTER` functions are used to count the number of users who left the website after visiting each page. The result is a table that shows the bounce rate for each page:

page_url	bounce_rate
`/homepage.html`	0.333333333
`/checkout.html`	0.666666667
`/products.html`	0.666666667
`/about.html`	1

Figure 9.39 – Result data

This table shows that the /about.html page has the highest bounce rate, meaning that all users who visited that page left the website without visiting any other pages. This suggests that there may be issues with the content or user experience on that page that need to be addressed.

Tips on when to use the Lead and Lag functions

Here are some tips on when to use the Lead and Lag functions:

- **Time-series analysis**: The Lead and Lag functions are often used in time-series analysis to calculate trends and changes in data over time. For example, you can use these functions to calculate the change in sales from one month to the next, or to identify patterns in website traffic over a period of time.

- **Cohort analysis**: Cohort analysis is a way of analyzing the behavior of a group of users over time. The Lead and Lag functions can be used to calculate the time between events for each user in a cohort, such as the time between the user's first and second purchases.

- **Comparative analysis**: The Lead and Lag functions can be used to compare the behavior of different groups of users. For example, you can use these functions to compare the time between events for users who made a purchase and users who did not make a purchase, or to compare the time between events for users who are in different age groups.

- **Query optimization**: The Lead and Lag functions can be used to optimize SQL queries. For example, you can use these functions to calculate the difference between two rows in a table without having to join the table to itself.

In general, the Lead and Lag functions are useful whenever you need to compare data between rows in a table or to analyze trends or changes in data over time.

NTILE()

The NTILE function in SQL is used to divide a result set into a specified number of equally sized groups or buckets. It assigns a rank to each row within the result set based on the specified number of buckets, with each row assigned to a bucket based on its rank.

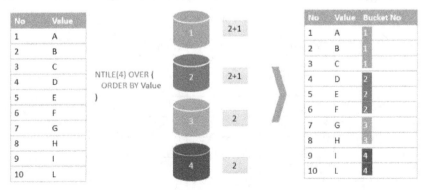

Figure 9.40 – NTILE function

The syntax for the NTILE function is as follows:

```
NTILE(n) OVER (ORDER BY column_name)
```

Here, n specifies the number of buckets to divide the result set into, and column_name specifies the column to order the result set by. The NTILE function returns an integer value that represents the bucket number to which the current row belongs.

For example, suppose we have a table called sales with the following columns:

sale_id	sale_date	Amount
1	1/1/2022	100
2	1/2/2022	200
3	1/3/2022	150
4	1/4/2022	300
5	1/5/2022	250
6	1/6/2022	175
7	1/7/2022	225
8	1/8/2022	175
9	1/9/2022	250
10	1/10/2022	200

Figure 9.41 – sales table

To divide this table into four equally sized buckets based on the amount of each sale, we could use the following query:

```
SELECT sale_id, amount, NTILE(4) OVER (ORDER BY amount) AS bucket
FROM sales;
```

This query assigns each row to a bucket based on the amount of the sale, with bucket 1 containing the smallest sales and bucket 4 containing the largest sales:

sale_id	amount	bucket
1	100	1
8	175	1
6	175	2
3	150	2
2	200	3
10	200	3
7	225	3

5	250	4
9	250	4
4	300	4

Figure 9.42 – Result data

The NTILE function can be useful for a variety of tasks, such as calculating quartiles, analyzing customer behavior, and performing A/B testing.

Scenario

Let's take the example of A/B testing to explain how to use the NTILE function in SQL. A/B testing is a common technique used in digital marketing and product development to test the effectiveness of different versions of a web page or application. In an A/B test, users are randomly assigned to one of two groups: the control group and the experimental group. The control group sees the original version of the web page, while the experimental group sees a modified version.

To analyze the results of an A/B test, we can use the NTILE function in SQL to divide the users into equal-sized buckets based on their behavior. For example, we could use the NTILE function to divide the users into four equally sized buckets based on the number of pages they visited on our website. We can then compare the behavior of the control group and the experimental group within each bucket.

Suppose we have a table called user_data with the following columns:

user_id	Group	pages_visited
1	A	2
2	B	3
3	A	4
4	B	1
5	A	6
6	B	2
7	A	3
8	B	4
9	A	5
10	B	2

Figure 9.43 – User data

We can use the following SQL query to divide the users into four buckets based on the number of pages they visited:

```
SELECT user_id, group, pages_visited, NTILE(4) OVER (ORDER BY pages_
visited) AS bucket
FROM user_data;
```

user_id	Group	pages_visited	Bucket
4	B	1	1
1	A	2	1
6	B	2	2
7	A	3	2
2	B	3	3
3	A	4	3
8	B	4	4
5	A	6	4
9	A	5	4
10	B	2	4

Figure 9.44 – Result data

We can then compare the behavior of the control group and the experimental group within each bucket. For example, we can calculate the average number of pages visited by the control group and the experimental group within each bucket, and then compare the two groups to see if the modified version of the web page had a significant effect on user behavior.

The NTILE function can be a useful tool for analyzing the results of an A/B test, as it allows us to divide users into equally sized buckets and compare the behavior of different groups within each bucket.

Now, let's see what we can infer from these results as a data analyst:

- **User IDs and groups**: Each row represents a user with a unique user ID and belongs to a specific group (**A** or **B**). This information helps identify and analyze user behavior based on their group membership.

- **Pages visited**: The **pages_visited** column indicates the number of pages visited by each user. Users with higher values in this column have visited more pages, indicating a higher level of engagement or activity on the platform.

- **Bucket**: The **bucket** column represents a categorization or segmentation of users based on their **pages_visited** value. Users with similar numbers of pages visited are grouped together in the same bucket.

- **Group comparison**: By examining the data, users can compare the behavior of users in Group A with those in Group B. They can analyze metrics such as average pages visited, distribution across buckets, or any other relevant patterns to understand the differences or similarities between the two groups.

- **User engagement**: The data allows users to evaluate the level of engagement or activity on the platform based on the number of pages visited. Users with higher page visit counts, especially those in Bucket 4, can be considered highly engaged users who are actively exploring various pages.

- **Bucket analysis**: Analyzing the distribution of users across buckets can provide insights into user segmentation based on their activity level. Users in Bucket 1 may be considered less active, while those in Bucket 4 are likely the most active and engaged users.

These interpretations can assist in understanding user behavior, identifying trends, and making data-driven decisions to optimize user experience and engagement on the platform.

The `NTILE` function in SQL can be useful in a variety of data-wrangling scenarios. Here are a few examples:

- **Market Segmentation**: `NTILE` can be used to segment customers or market data into equal-sized groups based on a particular variable, such as age, income, or spending behavior. This can help businesses identify different customer segments with distinct needs and preferences and tailor marketing or product offerings accordingly.

- **Ranking and Scoring**: `NTILE` can be used to rank and score data based on a particular variable, such as sales performance or customer satisfaction. This can help businesses identify top performers or areas for improvement and allocate resources more effectively.

- **Statistical Analysis**: `NTILE` can be used in statistical analysis to divide data into equal-sized groups for the purpose of calculating quartiles, percentiles, or other statistical measures. This can help businesses gain insights into the distribution of data and identify outliers or anomalies.

- **A/B Testing**: `NTILE` can be used in A/B testing to segment test groups into equal-sized buckets based on certain characteristics, such as demographics or behavior. This can help businesses compare the performance of different test groups and identify which variations are most effective.

Overall, the `NTILE` function in SQL is a versatile tool for data wrangling that can be applied to a wide range of scenarios and applications.

Summary

This brings us to the end of this chapter. Let's summarize the key topics that have been covered:

- Understand the concept of a window function, which is a calculation that operates on a window of rows within a table

- Identify common types of window functions, including aggregate and ranking functions, and be able to use them in SQL queries

- Understand the syntax of a window function, including the `OVER` clause, which is used to define the window of rows to operate on

- Use the PARTITION BY clause to group rows into partitions for calculation purposes

- Understand the difference between window functions and aggregate functions, which operate on groups of rows rather than windows of rows

- Understand the difference between window functions and subqueries, which can both be used to calculate values based on sets of rows, but operate in different ways

- Understand the limitations of window functions and how they may perform differently based on the underlying database technology

You should be able to use window functions to perform complex calculations on sets of rows within a table, and you should have a deeper understanding of how SQL queries can be used to manipulate and analyze data.

In the next chapter, we will learn how to become an effective coder. You will learn how to optimize SQL queries for better performance by analyzing query execution plans, identifying bottlenecks, using indexes effectively, rewriting queries, and improving database design. The goal is to improve query response times and reduce resource usage, leading to faster and more efficient data processing.

Part 4:
Optimizing Query Performance

This part includes the following chapter:

- *Chapter 10, Optimizing Query Performance*

Optimizing Query Performance

In this chapter, we will cover various techniques and strategies to improve the speed and efficiency of SQL queries. We will talk about query optimization and the importance of understanding query execution plans. Various key topics such as indexing, selecting appropriate data types, table partitioning, caching, normalization, and writing efficient SQL queries will be discussed. We will also cover advanced topics such as aggregations, query performance tuning techniques, monitoring, and troubleshooting slow-performing queries. This chapter will conclude by emphasizing the importance of continuous optimization to ensure that query performance remains optimized over time.

Introduction to query optimization

Query optimization is the process of improving the performance of a SQL query by reducing the time it takes to execute and retrieve the data. The goal of query optimization is to make the query run faster, consume fewer resources, and return the desired results with minimum overhead. Query optimization becomes critical when dealing with large amounts of data and when the data needs to be processed quickly:

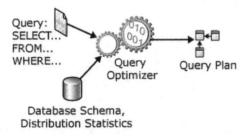

Figure 10.1 – Query optimization

In this section, we'll explain the importance of query optimization by mentioning the following points:

- **Scalability**: As the data grows, the queries may take longer to run, leading to slower performance and longer wait times for the users. Query optimization helps in ensuring that the queries scale well and continue to perform efficiently as the data grows.

- **Resource utilization**: Query optimization helps in reducing the resources required to run the query. This means lower CPU, memory, and disk usage, freeing up resources for other tasks.

- **User satisfaction**: Faster query performance leads to a better user experience as the users receive their results quickly. Query optimization helps in providing users with fast and reliable results, leading to higher user satisfaction.

- **Cost savings**: Query optimization can lead to cost savings as it reduces the number of resources required to run the query, including hardware, electricity, and maintenance costs.

- **Improved accuracy**: Query optimization can lead to improved accuracy as the optimized queries return the correct results faster and with fewer errors.

- **Better decision-making**: Faster and more accurate results lead to better decision-making as the data is available to decision-makers in real time, allowing them to make informed decisions based on the latest information.

- **Better performance for business-critical queries**: Query optimization is particularly important for business-critical queries as these queries often involve large amounts of data and are critical to the success of the business. Optimizing these queries can result in a significant impact on the overall performance of the business.

The rest of this chapter will build on this foundation by discussing the various techniques and strategies for optimizing SQL queries.

Query execution plan

Understanding query execution plans is an essential aspect of query optimization. It refers to the steps that the database takes to retrieve the data from the database and return the results to the user. Understanding these steps is critical in identifying bottlenecks and inefficiencies in the query, which can then be addressed to improve the performance of the query:

- **What is a query execution plan?** A query execution plan is a detailed map of the steps that the database takes to execute a SQL query. It shows the order in which the database accesses the data, the methods used to retrieve the data, and the resources required to run the query.

- **How can we view a query execution plan?** Query execution plans can be viewed using tools provided by the database management system, such as the EXPLAIN PLAN in Oracle or the SHOWPLAN in SQL Server. These tools provide a graphical representation of the query execution plan, making it easier to understand and analyze.

- **How can we interpret query execution plans?** Understanding how to interpret a query execution plan is crucial in identifying inefficiencies and bottlenecks in the query. For example, a large number of table scans may indicate that the query is not taking advantage of indexes, leading to slow performance.

- **How can we optimize query execution plans?** Once the inefficiencies and bottlenecks have been identified, they can be addressed so that we can optimize the query execution plan. This may involve adding indexes, changing the order of the tables in the join, or using different join methods to access the data.

In conclusion, understanding query execution plans is a crucial aspect of query optimization. By analyzing the steps that the database takes to execute the query, inefficiencies and bottlenecks can be identified and addressed to improve the performance of the query. The rest of this chapter builds on this foundation by discussing specific techniques and strategies for optimizing the query execution plan.

Query optimization techniques

When we write SQL queries, we're essentially asking the database to find specific data that meets certain criteria. Query optimization is the process of improving the performance of these queries. Essentially, we want our queries to be as fast and efficient as possible so that we can get the results we need quickly.

Here are a few common query optimization techniques:

- **Indexing**: When we create an index on a table, it helps the database find the data we're looking for more quickly. It's like creating a table of contents for a book – instead of searching through the whole book to find what we need, we can just look at the table of contents to find the relevant page. Just like how an index in a phone book helps you find information faster, an index in SQL helps the database find data more efficiently, resulting in faster query execution.

- **Joins**: When we need to combine data from multiple tables, we often use a join. However, joins can be expensive in terms of performance. To optimize queries that use joins, we can try to reduce the amount of data that needs to be joined, or we can use techniques such as subqueries to break the query down into smaller parts.

> **Note**
> Comparing strings is generally more computationally expensive than comparing integers. When joining tables on a string column, the database needs to perform string comparisons, which involve comparing each character in the strings. In contrast, integer comparisons are simpler and faster because they involve comparing numeric values directly.

- **Aggregation**: When we need to perform calculations on a large amount of data, it can be slow to do so all at once. Instead, we can use aggregation functions such as SUM, AVG, COUNT, and MAX to perform calculations on smaller subsets of the data.

- **Query structure**: The way we structure our queries can have a big impact on performance. For example, we can try to use more specific criteria to filter our data, rather than using generic criteria that will return a lot of results. We can also try to minimize the number of subqueries and the complexity of our queries.

These are just a few examples of query optimization techniques. Ultimately, the goal is to make our queries as fast and efficient as possible so that we can get the data we need without having to wait too long for it. Let's understand this via an example scenario.

Example

Let's consider a simplified version of Spotify's dataset, which includes two tables – users and songs. Here's what they might look like:

user_id	username	country
1	john123	USA
2	sarah456	UK
3	lisa789	USA
4	mike012	Canada

Figure 10.2 – Users table

song_id	song_name	artist	length
1	Hello	Adele	4:55
2	Shape of You	Ed Sheeran	3:54
3	Lose You To Love Me	Selena Gomez	3:27
4	Blinding Lights	The Weeknd	3:21

Figure 10.3 – Songs table

Now, let's say we want to find all the songs that were played by users from the USA. We can do this using a SQL query, like this:

```
SELECT songs.song_name, songs.artist
FROM songs
JOIN (
  SELECT user_id
  FROM users
  WHERE country = 'USA'
) AS us_users
ON songs.user_id = us_users.user_id;
```

However, this query might not be very efficient, especially if the users and songs tables are very large. Here are a few query optimization techniques we could use to make this query faster:

- **Indexing**: We could create an index on the country column in the users table, which would make it faster to filter by country.

- **Query structure**: We could simplify the query by removing the subquery and using a regular join instead. This would make the query easier for the database to execute since it wouldn't need to process a separate subquery. Here's what the simplified query would look like:

```
SELECT songs.song_name, songs.artist
FROM songs
JOIN users
ON songs.user_id = users.user_id
WHERE users.country = 'USA';
```

- **Aggregation**: We could use aggregation to group the results by song and artist, and then count the number of times each song was played. This would give us a sense of which songs were most popular among users from the USA. Here's what the aggregated query would look like:

```
SELECT songs.song_name, songs.artist, COUNT(*) as num_plays
FROM songs
JOIN users
ON songs.user_id = users.user_id
WHERE users.country = 'USA'
GROUP BY songs.song_name, songs.artist
ORDER BY num_plays DESC;
```

These are just a few examples of query optimization techniques that could be used to make our queries faster and more efficient. By using these techniques, we can reduce the amount of time it takes to get the data we need and make our applications more responsive and scalable.

Table partitioning

Table partitioning is a technique that's used to improve the performance of large tables by dividing them into smaller, more manageable pieces. Imagine you have a table that contains millions of rows of data. When you want to query this table, it can take a long time for the database to scan through all those rows to find the data you need. Table partitioning solves this problem by dividing the table into smaller chunks, based on some criteria such as date, geographical location, or some other column in the table. This allows the database to scan only the relevant partitions instead of the entire table, making queries faster and more efficient:

Example

Let's say that Instagram Reels has a large table of user data that includes information about each user's activity on the app. The table contains millions of rows, which can make queries slow and inefficient. To improve performance, Instagram Reels could partition the user data table by date. Here's an example of how the user data table might be partitioned in SQL:

```
CREATE TABLE user_data (
   user_id INT,
   activity_date DATE,
   activity_type VARCHAR(50),
   activity_count INT
)
PARTITION BY RANGE (activity_date) (
   PARTITION p201901 VALUES LESS THAN ('2019-02-01'),
   PARTITION p201902 VALUES LESS THAN ('2019-03-01'),
   PARTITION p201903 VALUES LESS THAN ('2019-04 01'),
   PARTITION p201904 VALUES LESS THAN ('2019-05-01'),
   ...
);
```

In this example, the user_data table has been partitioned by the activity_date column using the RANGE partitioning method. The table has been split into partitions based on the value of the activity_date column, with each partition containing data for a specific date range.

Caching

Caching is a technique that's used to speed up the retrieval of data by storing frequently accessed data in memory, rather than retrieving it from the database every time it's needed. Caching is like having a bookmark for frequently accessed data so that you can quickly access it without having to search for it every time.

Let's say that Instagram Reels has a feature that displays a user's most recent videos in a "Recent Activity" section of the app. To make this feature fast and responsive, Instagram Reels could cache the user's most recent videos in memory. Here's an example of how caching might be implemented in SQL:

```
SELECT * FROM user_videos
WHERE user_id = 12345
ORDER BY video_date DESC
LIMIT 10;
```

In this example, the SQL query retrieves the 10 most recent videos for a specific user. To improve performance, Instagram Reels could cache the results of this query in memory so that the next time the query is executed, the results can be retrieved quickly from the cache without it hitting the database.

Normalization

Normalization is a process that's used to organize data in a database so that it's more efficient and easier to maintain. Normalization involves breaking up large tables into smaller tables and establishing relationships between them. The goal is to reduce data redundancy and improve data consistency.

Here's an example of how normalization might be implemented in SQL:

```
CREATE TABLE users (
  user_id INT,
  username VARCHAR(50),
  email VARCHAR(50)
);
CREATE TABLE user_profiles (
  user_id INT,
  profile_picture VARCHAR(255),
  bio VARCHAR(255)
);
```

In this example, the user data is split into two tables: `users` and `user_profiles`. The `users` table contains information such as the user's ID, username, and email address, while the `user_profiles` table contains information such as the user's profile picture and bio. The two tables are linked together by the `user_id` column, which establishes a relationship between the tables. This reduces data redundancy and makes it easier to maintain data consistency since changes to user information only need to be made in one place.

Query monitoring and troubleshooting

Query monitoring and troubleshooting involve identifying slow-performing queries and finding ways to improve their performance. Let's look at some techniques for monitoring and troubleshooting slow queries.

Query profiling

Query profiling is the process of analyzing the query execution plan to identify the steps that are taking the most time. This can help you identify the parts of the query that need to be optimized:

Student ID	Course ID	Grade
1	101	85
2	101	92
3	101	76
1	102	90
2	102	82
3	102	88

Figure 10.4 – Grades table

To illustrate query profiling, let's say we want to analyze the performance of a query that calculates the average grade for a given course. We might write a query like this:

```
SELECT AVG(Grade) FROM grades WHERE CourseID = 101;
```

To profile this query, we can use a tool such as SQL Server Profiler, which allows us to capture the SQL statements that are executed, along with their execution plans and performance metrics:

```
|--Compute Scalar(DEFINE:([Expr1003]=CASE WHEN [Expr1004]=(0) THEN
NULL ELSE [Expr1002] / [Expr1004] END))
   |--Stream Aggregate(DEFINE:([Expr1002]=SUM([tempdb].[dbo].[grades].
[Grade]), [Expr1004]=COUNT_BIG(*)))
      |--Clustered Index Scan(OBJECT:([tempdb].[dbo].[grades].
[PK__grades__CF5685BE6B41D7B3]), WHERE:([tempdb].[dbo].[grades].
[CourseID]=(101)))
```

This execution plan shows that the database engine is performing a clustered index scan on the grades table to find the rows where CourseID = 101. Then, it is performing a stream aggregate to compute the sum and count of the grades for that course, and finally, a compute scalar to compute the average grade.

We can also capture performance metrics such as the duration of the query, the number of reads and writes, and the number of rows returned. By analyzing these metrics, we can identify which parts of the query are taking the most time, and find ways to optimize the query.

To illustrate query logging, let's say we want to log all the queries that are executed against the `grades` table. We might create a simple logging table, like this:

```
CREATE TABLE query_log (
    QueryText NVARCHAR(MAX),
    ExecutionTime DATETIME,
    Duration INT,
    RowsAffected INT
);
```

Then, we can use a trigger to log all the queries that are executed:

```
CREATE TRIGGER log_query
ON grades
AFTER INSERT, UPDATE, DELETE
AS
BEGIN
    INSERT INTO query_log (QueryText, ExecutionTime, Duration,
RowsAffected)
    SELECT
        EVENTDATA().value('(/EVENT_INSTANCE/TSQLCommand/CommandText)
[1]', 'NVARCHAR(MAX)'),
        GETDATE(),
        DATEDIFF(ms, EVENTDATA().value('(/EVENT_INSTANCE/StartTime)
[1]', 'DATETIME'), EVENTDATA().value('(/EVENT_INSTANCE/EndTime)[1]',
'DATETIME')),
        @@ROWCOUNT;
END;
```

This trigger will log the text of the executed query, the time it was executed, the duration of the query, and the number of rows affected.

By analyzing the query log, we can identify slow-performing queries, find ways to optimize them, and track changes in query performance over time. We can also use the log to audit the queries that are executed and identify any security issues or performance problems.

Query logging

Query logging is the process of recording all the queries that are executed in the database, along with their execution time and other relevant information. This can help you identify slow-performing queries and find ways to optimize them.

Let's use the same example dataset of a university grade system. Suppose we have a grades table that contains the following data:

Student ID	Course ID	Grade
1	101	85
2	101	92
3	101	76
1	102	90
2	102	82
3	102	88

Figure 10.5 – Grades table

To enable query logging, we can create a new table named `query_log` that consists of the following columns:

```
CREATE TABLE query_log (
    query_text NVARCHAR(MAX),
    execution_time DATETIME,
    duration INT,
    rows_affected INT
);
```

Then, we can create a trigger on the `grades` table that logs information about each query that is executed:

```
CREATE TRIGGER log_query
ON grades
AFTER INSERT, UPDATE, DELETE
AS
BEGIN
    DECLARE @query_text NVARCHAR(MAX);
    SET @query_text = (SELECT TEXT FROM sys.dm_exec_sql_text(SQL_
HANDLE));

    INSERT INTO query_log (query_text, execution_time, duration, rows_
affected)
    VALUES (
        @query_text,
        GETDATE(),
        DATEDIFF(ms, sys.dm_exec_requests.start_time, sys.dm_exec_
requests.end_time),
```

```
        @@ROWCOUNT
    );
END;
```

Now, every time a query is executed on the `grades` table, information about the query is logged in the `query_log` table. For example, let's say we execute a query to update a student's grade, like this:

```
UPDATE grades SET Grade = 95 WHERE StudentID = 1 AND CourseID = 101;
```

The trigger we created will log information about this query in the `query_log` table. The log might look something like this:

query_text	execution_time	duration	rows_affected
UPDATE grades SET Grade = 95 WHERE ...	2023-02-17 10:30:15.34	15	1

Figure 10.6 – Query log table

This log shows us the text of the query that was executed, the time it was executed, how long it took to run, and how many rows were affected. We can use this information to monitor the queries that are executed on the `grades` table, identify slow-performing queries, and diagnose any performance issues.

Database monitoring

Database monitoring is the practice of keeping an eye on the health and performance of a database to ensure that it is running smoothly. It involves tracking various metrics and events related to the database, such as its size, throughput, response time, and availability.

To monitor a database, you might use monitoring tools that help you collect and analyze data about the database. These tools might provide graphs and charts that allow you to visualize the performance of the database over time. They might also send alerts if certain performance thresholds are exceeded, such as if the database is taking too long to respond to queries or if there is a sudden increase in the number of errors being reported.

Some of the key metrics that you might monitor in a database include the following:

- **Throughput**: This measures how many requests the database is handling per unit of time. For example, you might monitor the number of transactions per second or the number of queries per minute.
- **Response time**: This measures how quickly the database is responding to requests. You might monitor the average response time for various types of queries or transactions.

- **Size**: This measures the amount of data stored in the database. You might monitor the total size of the database or the size of individual tables.

- **Availability**: This measures the percentage of time that the database is up and running. You might monitor downtime or outages to ensure that the database is available when it's needed.

Overall, database monitoring is an important practice for ensuring the smooth operation of a database. By keeping a close eye on key metrics and events, you can identify performance issues early and take steps to address them before they become major problems. By using these techniques, you can identify and optimize slow-performing queries to improve the overall performance of your database system.

Tips and tricks for writing efficient queries

As a data citizen, it is key to write an efficient query to ensure all the resources are being utilized correctly. Here are some tips and tricks for writing efficient SQL queries for data wrangling:

- **Be selective**: Only select the columns you need. Selecting all columns in a table when you only need a few can be a major performance bottleneck. By limiting your selection to the necessary columns, you can reduce the amount of data that needs to be transferred and processed.

- **Use indexes**: Indexes can greatly improve query performance by allowing the database to quickly locate the data it needs. Make sure you create indexes on the columns that are frequently used in your queries.

- **Avoid subqueries**: Subqueries can be very slow, especially if they are nested. Whenever possible, try to avoid using subqueries and use join operations instead.

- **Use WHERE clauses**: The WHERE clause can help filter data before it is processed, which can significantly improve performance. When using the WHERE clause, make sure you use appropriate comparison operators (for example, =, <, and >) and avoid using functions or calculations.

- **Avoid using DISTINCT**: Using the DISTINCT keyword to eliminate duplicates can be very slow, especially on large datasets. Instead, use GROUP BY clauses to group data and eliminate duplicates.

- **Use UNION ALL**: When combining multiple tables, use the UNION ALL operator instead of the UNION operator. UNION ALL is faster because it does not remove duplicates.

- **Use appropriate joins**: Make sure you use the appropriate type of join for the data you are working with. Inner joins are generally faster than outer joins, but outer joins may be necessary in certain situations.

- **Consider denormalizing data**: In some cases, denormalizing data (that is, combining multiple tables into a single table) can improve query performance. However, this should only be done if it will result in a significant performance improvement and does not compromise data integrity:

Tips for SQL Query Optimization

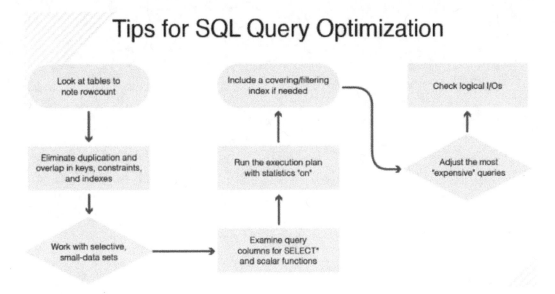

Figure 10.7 – SQL query optimization flowchart

Overall, writing efficient SQL queries for data wrangling requires a combination of knowledge about the data you are working with and the SQL language. By following these tips and tricks, you can improve the performance of your queries and save time when working with large datasets.

Summary

This brings us to the end of this chapter. We are sure you have learned and taken key notes that will be helpful in your day-to-day work. At this point, we have learned about different methods within SQL queries that cause delays in data transformation. We also know the key difference between a good query and a bad query and how each produces the same result with different efficiencies.

In the next chapter, we will learn about descriptive statistics using SQL, which will provide us with insights into the distribution, central tendency, and variability of data, which can, in turn, help us identify outliers and anomalies. Common SQL functions and statements used for descriptive statistics include COUNT, AVG, MIN, MAX, and GROUP BY. By using SQL to analyze data, researchers and analysts can efficiently extract and summarize information from large datasets.

Part 5:
Data Science And Wrangling

This part includes the following chapters:

11
Descriptive Statistics with SQL

Descriptive statistics is a fundamental aspect of data analysis that helps us to summarize and describe the main characteristics of a dataset. With the increasing availability of large datasets, it has become more important than ever to have tools and techniques to help us understand the data we are working with.

In this chapter, we will explore how to use SQL to calculate various descriptive statistics measures, such as mean, median, mode, standard deviation, and variance. We will also demonstrate how to generate visualizations, such as histograms and box plots, to gain insights into the distribution of data.

Throughout the chapter, we will use real-world examples to demonstrate the application of SQL in descriptive statistics. We will assume that you have a basic understanding of SQL and statistical concepts such as mean, median, and standard deviation.

By the end of this chapter, you will have a solid understanding of how to use SQL to calculate and interpret descriptive statistics and be able to apply these techniques to your own datasets.

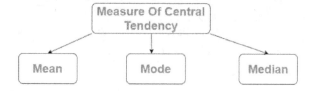

Figure 11.1 – Measure of central tendency

Calculating descriptive statistics with SQL

Calculating descriptive statistics with SQL is an important topic in data analysis, as it allows us to summarize and understand the main characteristics of a dataset. Here are some examples of how to calculate various measures of central tendency and variability in SQL.

Mean

The mean, also known as the average, is a measure of central tendency that represents the sum of all values in a dataset divided by the number of observations. In SQL, we can calculate the mean using the AVG() function. For example, to calculate the mean of the salary column in the employees table, we can use the following query:

```
SELECT AVG(salary) AS mean_salary
FROM employees;
```

Case scenario

An interesting real-world scenario for calculating the mean in SQL for descriptive statistics is to analyze the average time spent by visitors on a website. Assume we have a table named pageviews containing records of page views with columns such as visitor_id, page_url, page_title, page_timestamp, and time_on_page.

visitor_id	page_url	page_title	page_timestamp	time_on_page
1	/home	Home Page	3/1/2022 8:00	120
2	/about	About Us	3/1/2022 8:01	60
3	/product/1234	Product Page	3/1/2022 8:02	240
4	/product/5678	Another Product	3/1/2022 8:04	180
5	/home	Home Page	3/1/2022 8:07	300
6	/product/1234	Product Page	3/1/2022 8:09	90
7	/about	About Us	3/1/2022 8:11	120
8	/product/5678	Another Product	3/1/2022 8:13	150
9	/home	Home Page	3/1/2022 8:16	420
10	/product/1234	Product Page	3/1/2022 8:22	210
11	/product/5678	Another Product	3/1/2022 8:27	90
12	/home	Home Page	3/1/2022 8:31	180

Figure 11.2 – pageviews table

To calculate the mean time spent by visitors on the website, we can use the AVG() function in SQL as follows:

```
SELECT AVG(time_on_page) as avg_time_on_page
FROM pageviews;
```

Here's the output of the query based on the sample data provided:

```
mean_time_on_page
185.4545
```

So, the mean time spent on each page based on the given data is approximately 185.4545 seconds.

This query would return the average time spent by visitors on all the pages in the `pageviews` table. By analyzing this data, website owners could gain insights into how engaging their website is and identify which pages have the highest bounce rates. They could also use this information to optimize their website's user experience and improve their website's overall engagement metrics. For example, they could identify which pages have the highest "time on page" and determine which elements on the page are contributing to longer visitor engagement. They could then incorporate similar elements into other pages to improve engagement across the site.

Interesting read

An interesting industry example of how to determine the mean in SQL for descriptive statistics is to calculate the average daily sales for a retail store chain. Assume we have a table named `sales` containing daily sales records with columns such as `store_id`, `date`, `total_sales_amount`, and `customer_count`.

To calculate the average daily sales for the retail store chain, we can use the `AVG()` function in SQL as follows:

```
SELECT AVG(total_sales_amount) as avg_daily_sales
FROM sales;
```

This query would return the average daily sales for all the stores in the `sales` table. The retail store chain could use this information to analyze the performance of their stores, identify the best and worst performing stores, determine the effectiveness of their marketing and sales strategies, and make data-driven decisions to improve their sales performance. For example, they could use this information to adjust their inventory levels, optimize their pricing strategies, and allocate resources more effectively to achieve higher sales and profitability.

Median

The median is the middle value of a dataset when the values are arranged in order. In SQL, we can calculate the median using the `PERCENTILE_CONT()` function. For example, to calculate the median of the `age` column in the `customers` table, we can use the following query:

```
SELECT PERCENTILE_CONT(0.5) WITHIN GROUP (ORDER BY age) AS median_age
FROM customers;
```

Interesting read

Let's look at an interesting real-world scenario for calculating the median in SQL for descriptive statistics.

Suppose you are a data analyst for a retail company that sells clothing online. You are interested in understanding the distribution of purchase amounts for a specific product category, in order to inform pricing and inventory decisions. You have access to a `purchases` table that includes information on all customer purchases for this product category, including `purchase_date`, `customer_id`, and `purchase_amount`.

purchase_date	customer_id	product_category	purchase_amount
1/1/2022	123	Clothing	50
1/2/2022	456	Clothing	75
1/3/2022	789	Clothing	120
1/4/2022	123	Clothing	40
1/5/2022	456	Clothing	90
1/6/2022	789	Clothing	65
1/7/2022	123	Clothing	55

Figure 11.3 – purchases table

You want to calculate the median purchase amount for this product category, as it provides a useful measure of the central tendency of the data that is not influenced by extreme outliers.

To do this in SQL, you would write a query that selects all the purchase amounts from the `purchases` table, orders them in ascending order, and then selects the middle value as the median. Here's an example query:

```
SELECT
    AVG(purchase_amount) AS median_purchase_amount
FROM (
    SELECT
        purchase_amount,
        ROW_NUMBER() OVER (ORDER BY purchase_amount) AS row_num,
        COUNT(*) OVER() AS total_rows
    FROM
        purchases
    WHERE
        product_category = 'clothing'
) AS t
WHERE
    row_num IN (FLOOR((total_rows + 1) / 2), CEIL((total_rows + 1) /
2));
```

This query uses a subquery to calculate the row number for each purchase amount in the ordered list, as well as the total number of rows in the table. It then selects the average of the two middle values as the median.

The resulting output would be a single row with a column named median_purchase_amount, which would contain the calculated median value for the purchase amounts in the purchases table for the specified product category. This metric could then be used to inform pricing and inventory decisions for this product category.

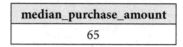

median_purchase_amount
65

Figure 11.4 – output table

The output table shows a single row with the calculated median purchase amount of **65.00** for the purchases table for the specified product category.

Mode

The mode is the value that appears most frequently in a dataset. In SQL, we can calculate the mode using the MODE() function. For example, to calculate the mode of the color column in the products table, we can use the following query:

```
SELECT MODE() WITHIN GROUP (ORDER BY color) AS mode_color
FROM products;
```

Interesting read

Let's look at an interesting real-world scenario for calculating the mode in SQL for descriptive statistics.

Suppose you work for a market research company that conducts surveys to understand consumer preferences for various products. You have a table called survey_results that contains the responses to a survey question asking consumers which brand of cola they prefer. The survey_results table has a single column called brand.

Here's an example SQL query to find the mode brand from the survey_results table:

```
SELECT brand, COUNT(*) as count
FROM survey_results
GROUP BY brand
ORDER BY count DESC
LIMIT 1;
```

This query groups the rows in the survey_results table by brand, counts the number of responses for each brand, sorts the results in descending order by count, and then selects the top row as the mode.

For example, suppose the `survey_results` table looks like this:

brand
Pepsi
Coke
Pepsi
RC Cola
Coke
Pepsi
Pepsi

Figure 11.5 – survey results table

The output of the SQL query would be a single row with two columns, `brand` and `count`, which would contain the calculated mode value for the brands in the `survey_results` table:

brand	count
Pepsi	4

Figure 11.6 – output table

In this case, the mode brand preferred by the survey respondents is Pepsi, which can be a useful measure of the most popular brand among consumers.

Standard deviation

The standard deviation is a measure of variability that represents how spread out the values in a dataset are from the mean. In SQL, we can calculate the standard deviation using the `STDDEV()` function. For example, to calculate the standard deviation of the `weight` column in the `products` table, we can use the following query:

```
SELECT STDDEV(weight) AS std_weight
FROM products;
```

An industry example of how to use SQL for descriptive statistics is to calculate the standard deviation of employee salaries in a company. Assume we have a table named `employees` containing employee records with columns such as `employee_id`, `first_name`, `last_name`, `job_title`, `department_name`, and `salary`.

To calculate the standard deviation of employee salaries, we can use the STDDEV() function in SQL as follows:

```
SELECT STDDEV(salary) as salary_std_deviation
FROM employees;
```

This query would return the standard deviation of salaries for all employees in the employees table. The company could use this information to determine the level of salary variability across different departments or job titles, identify employees who might be overpaid or underpaid, and make data-driven decisions regarding compensation and benefits.

Case scenario

Let's look at an interesting real-world scenario for calculating standard deviation in SQL for descriptive statistics.

Suppose a company wants to measure the performance of its sales team based on the number of sales made by each team member in a given period. They have a table named sales that contains the following columns: sales_id, team_member_id, sale_date, and sale_amount.

sales_id	team_member_id	sale_date	sale_amount
1	1	1/1/2022	10000.00
2	1	1/2/2022	5000.00
3	1	1/3/2022	7500.00
4	2	1/1/2022	8000.00
5	2	1/2/2022	7000.00
6	2	1/3/2022	9000.00
7	3	1/1/2022	12000.00
8	3	1/2/2022	5000.00
9	3	1/3/2022	8000.00
10	4	1/1/2022	7000.00
11	4	1/2/2022	6000.00
12	4	1/3/2022	9000.00
13	5	1/1/2022	5000.00
14	5	1/2/2022	4000.00
15	5	1/3/2022	3000.00

Figure 11.7 – sales table

To analyze the performance of the sales team, they want to calculate the standard deviation of the sale amount for each team member. This will help them to identify which team members are consistently performing well, and which ones are more unpredictable in their sales performance.

Here's an example query to calculate the standard deviation of the sale amount for each team member:

```
SELECT team_member_id, STDDEV(sale_amount) AS std_sale_amount
FROM sales
GROUP BY team_member_id;
```

This query calculates the standard deviation of `sale_amount` for each unique `team_member_id` in the `sales` table. The `GROUP BY` clause ensures that the results are grouped by `team_member_id`.

The output of the query will show the standard deviation of the sale amount for each team member, which will help the company identify which team members are performing consistently and which ones are more unpredictable in their sales performance. They can then use this information to provide additional training or support to those team members who are struggling to meet their targets.

Here's an example output of the query:

team_member_id	std_sale_amount
1	2500
2	1500
3	3000
4	2000
5	1000

Figure 11.8 – Output

In this example, the team member with `team_member_id` 5 has the lowest standard deviation of sale amounts, indicating that they are consistently performing well. The team member with `team_member_id` 3 has the highest standard deviation of sale amounts, indicating that their performance is more unpredictable.

Variance

The variance is another measure of variability that represents the average squared distance from the mean. In SQL, we can calculate the variance using the `VARIANCE()` function. For example, to calculate the variance of the `sales` column in the `sales_data` table, we can use the following query:

```
SELECT VARIANCE(sales) AS var_sales
FROM sales_data;
```

By calculating these measures of central tendency and variability in SQL, we can gain insights into the distribution and characteristics of a dataset, which can inform decision-making and further analysis.

Variability

Variability in descriptive statistics refers to the amount of dispersion or spread of a dataset around its central tendency. It is used to measure how much the values in a dataset vary or differ from one another.

One way to calculate variability in SQL is to use the variance function. The variance function calculates the average of the squared differences of each value in the dataset from the mean of the dataset. The result is a measure of the dispersion or variability of the data.

Another way to calculate variability in SQL is to use the standard deviation function. The standard deviation function is the square root of the variance function and is used to measure the degree of spread or dispersion of the data around the mean.

Both variance and standard deviation can be useful in a variety of fields, such as finance, science, and engineering. In finance, variability measures such as variance and standard deviation are used to assess the risk of investments, while in science and engineering, they can be used to evaluate the accuracy of measurements and to identify trends or patterns in data.

Case scenario

One interesting industry example of how variability can be used in SQL for descriptive statistics is in the financial industry, particularly for analyzing stock prices.

Stock prices are subject to variability due to a variety of factors, such as market sentiment, economic indicators, company news, and more. To better understand this variability, financial analysts can use SQL to calculate and analyze measures of variability such as variance.

For example, suppose a financial analyst wants to analyze the variability of a particular stock's price over a certain time period. They can use SQL to calculate the stock's daily returns and then calculate the standard deviation of those returns as a measure of volatility. They can also use SQL to calculate the variance of the stock's returns, which can provide additional insights into the distribution of returns.

By analyzing measures of variability in this way, financial analysts can gain a better understanding of the risk and uncertainty associated with a particular stock and make more informed investment decisions.

Let's say we have a table called `stock_prices` that contains the daily closing prices of a particular stock over a period of time:

stock_name	price	date
Apple	130.21	5/10/2021
Apple	131.23	5/11/2021
Apple	129.12	5/12/2021
Google	2422.52	5/10/2021
Google	2434.09	5/11/2021
Google	2418.49	5/12/2021
Microsoft	245.31	5/10/2021
Microsoft	248.08	5/11/2021
Microsoft	243.85	5/12/2021

Figure 11.9 – Stock prices table

This query calculates the variance of the `price` column for each `stock_name` group in the `stock_prices` table:

```
SELECT stock_name, VARIANCE(price) AS variance_price
FROM stock_prices
GROUP BY stock_name
```

The output will look like this:

stock_name	variance_price
Apple	1.091333333
Google	76.92626667
Microsoft	5.173333333

Figure 11.10 – Output

Calculating variance on a dataset can provide insights into the level of dispersion or variability in the data. In the case of stock prices, variance can be used to measure the degree of risk or volatility associated with investing in a particular stock.

A higher variance indicates that the stock prices are more spread out from the mean, and therefore, there is a greater degree of risk associated with investing in that stock. On the other hand, a lower variance indicates that the stock prices are relatively stable and there is a lower degree of risk associated with investing in that stock.

Investors can use the variance KPI to compare the risk levels of different stocks in their portfolio or to evaluate the risk-return trade-off for different investment options. Additionally, financial analysts may use variance as a metric to assess the overall market risk or to identify stocks that are more sensitive to market fluctuations.

Interesting read

A company that sells a variety of products is interested in analyzing their sales data to understand the performance of their products. They want to calculate the mean, median, and mode of sales revenue for each product category and also determine the variability in sales revenue.

Here's an example SQL query that can be used to calculate the mean, median, mode, and variability of sales revenue for each product category:

```
SELECT
    category,
    AVG(revenue) AS mean_sales,
    price AS median_sales,
    MODE() WITHIN GROUP (ORDER BY revenue) AS mode_sales,
    VARIANCE(revenue) AS variability
FROM sales_data
GROUP BY category
```

Let's say the `sales_data` table has the following columns: `product_name`, `category`, `price`, `units_sold`, and `revenue`.

Here's a sample output table that the preceding query might generate:

category	mean_sales	median_sales	mode_sales	variability
Electronics	50000	40000	35000	4000000
Clothing	30000	25000	20000	2000000
Beauty	20000	15000	10000	1000000

Figure 11.11 – Output

The `mean_sales` column shows the average sales revenue for each product category, which can be used to compare the overall performance of each category. For example, **Electronics** has the highest mean sales revenue of $50,000, indicating that it is the most profitable category.

The `median_sales` column shows the median sales revenue for each product category, which is the middle value when the sales revenue values are sorted in ascending order. It can be used to identify the typical sales revenue for each category. For example, **Clothing** has a median sales revenue of $25,000, indicating that half of the clothing products sold generated less than $25,000 in revenue.

The mode_sales column shows the mode sales revenue for each product category, which is the most frequently occurring value in the sales revenue distribution. It can be used to identify the most common sales revenue for each category. For example, in the **Electronics** category, the mode sales revenue is $35,000, indicating that this is the most common sales revenue generated by electronics products.

Finally, the variability column shows the variance of sales revenue for each product category, which can be used to determine how much the sales revenue values deviate from the mean. A higher variability indicates a wider spread of sales revenue values, which may be a cause for concern for the company. In this example, **Electronics** has the highest variability, $4,000,000, indicating that there is a wide range of sales revenue values for electronics products.

Summary

In conclusion, descriptive statistics using SQL is a powerful tool for analyzing and summarizing data. It allows businesses to extract useful information from large datasets, enabling better decision-making. In this chapter, we have covered various descriptive statistics, such as mean, median, mode, and variability and learned how to calculate them using SQL. We have also seen several examples of how these statistics can be used in real-world scenarios such as stock prices, survey data, and sales data.

Some key takeaways from this chapter include the following:

- Descriptive statistics are useful for summarizing and analyzing data
- SQL is a powerful tool for calculating descriptive statistics
- Mean, median, and mode are measures of central tendency, while variability measures the spread of the data
- Different descriptive statistics can provide different insights into data
- It is important to choose the appropriate descriptive statistics based on the nature of the data and the research question

Overall, descriptive statistics using SQL is a valuable technique that can help businesses gain insights into their data, identify patterns, and make informed decisions.

In the next chapter, we will learn how SQL can be used for time series analysis.

12
Time Series with SQL

A time series is a series of data points indexed in time order. Most commonly, a time series is a sequence taken at successive equally spaced points in time. Thus, it is a sequence of discrete-time data. Time series data is a sequence of data points. Each of the data points includes a timestamp. Timestamps usually include a date and then a time. Time series analysis comprises methods for analyzing time series data in order to extract meaningful statistics and other characteristics of the data. Time series forecasting is the use of a model to predict future values based on previously observed values. While regression analysis is often employed in such a way as to test relationships between one or more different time series, this type of analysis is not usually called "time series analysis," which refers to relationships between different points in time within a single series.

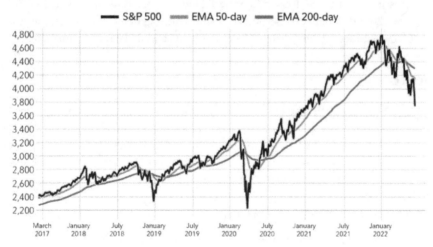

Figure 12.1 – Time series using SQL

Time series analysis is used in the following industrial areas:

- Stock market data
- Tide time tables
- Performance monitoring
- Health monitoring
- Population statistics
- Business performance

Time series analysis is a method of analyzing data that is collected over time. It involves looking at patterns and trends in the data to help identify potential relationships or forecast future outcomes.

In this chapter, the reader will learn how to use SQL to perform time series analysis. This means that they will learn how to use SQL queries to extract data from a database that has been collected over time. They will then be able to use different techniques and functions in SQL to analyze the data and uncover patterns or trends. For example, they might use SQL to extract data on sales for a particular product over the past year. They could then use time series analysis to identify trends, such as whether sales tend to be higher in certain months or whether there is a correlation between sales and advertising spend. They could also use forecasting techniques to predict future sales based on historical data.

Overall, time series analysis using SQL can be a powerful tool for businesses and researchers who want to better understand trends and patterns in their data. By learning these techniques, the reader will be able to use SQL to extract valuable insights from their data and make more informed decisions.

Running totals

Running totals, also known as cumulative totals, are a way of keeping track of a running sum of values as new data is added. In other words, it is a calculation that shows the total of a set of values up to a certain point in time. For example, let's say you have a table of daily sales for a store. You can use a running total to calculate the total sales up to each day of the week. This means that for Monday, the running total would be the total sales for Monday, and for Tuesday, the running total would be the total sales for Monday and Tuesday. The running-total calculation adds up the sales for each day and displays the total up to the current day. It's a useful tool for analyzing data over time and understanding trends.

In SQL, you can use a window function to calculate running totals. A window function allows you to perform calculations over a subset of rows, including running totals. By using the SUM function and the OVER clause in SQL, you can calculate the running total for a particular column. Overall, running totals are a simple but powerful tool for tracking cumulative data over time and can be useful in a variety of applications, such as financial analysis or sales tracking.

Case scenario

Let's say we have a table of daily website traffic data, where each row represents the number of visits to the website on a particular day. The table might look something like this:

Date	Visits
1/1/2022	1000
1/2/2022	1500
1/3/2022	1200
1/4/2022	1800
1/5/2022	2000
1/6/2022	1750
1/7/2022	2200

Figure 12.2 – Website traffic table

```
SELECT
  Date,
  SUM(Visits) OVER (ORDER BY Date) AS RunningTotal
FROM
  WebsiteTraffic;
```

In this query, the SUM function calculates the total website traffic for each day, and the OVER clause specifies that the calculation should be performed over a window of rows, which is ordered by date. The resulting table would look like this:

Date	RunningTotal
1/1/2022	1000
1/2/2022	2500
1/3/2022	3700
1/4/2022	5500
1/5/2022	7500
1/6/2022	9250
1/7/2022	11450

Figure 12.3 – Output table

As you can see, the RunningTotal column shows the cumulative total of website traffic up to each day. This can be a useful way to track website traffic trends and identify patterns over time. You can then use this data to generate **key performance indicators** (**KPIs**), such as total website traffic for a given time period, average daily website traffic, website traffic growth, and website traffic trends.

Lead and lag for time series analysis

As we have already read in the previous chapters, LEAD and LAG functions are used to refer to the previous or next row in the dataset for a specific column. Let's start understanding this with an example.

Case scenario

Imagine you have a table that tracks the number of sales made by a company each month, as follows:

Month	Sales
1/1/2021	100
2/1/2021	150
3/1/2021	200
4/1/2021	250
5/1/2021	300
6/1/2021	350
7/1/2021	400
8/1/2021	450
9/1/2021	500
10/1/2021	550
11/1/2021	600
12/1/2021	650

Figure 12.4 – Sales table

In time series analysis, you may want to analyze the changes in sales from one month to the next. This is where the LEAD and LAG functions come in.

LAG allows you to retrieve the value of a column from the previous row in a result set. In this case, you can use LAG to get the sales value from the previous month:

```
SELECT
    Month,
    Sales,
    LAG(Sales) OVER (ORDER BY Month) AS PrevSales
FROM
    SalesTable;
```

This query will return a table with three columns: Month, Sales, and PrevSales. The PrevSales column shows the sales value from the previous month:

Month	Sales	PrevSales
1/1/2021	100	NULL
2/1/2021	150	100
3/1/2021	200	150
4/1/2021	250	200
5/1/2021	300	250
6/1/2021	350	300
7/1/2021	400	350
8/1/2021	450	400
9/1/2021	500	450
10/1/2021	550	500
11/1/2021	600	550
12/1/2021	650	600

Figure 12.5 – Output table

On the other hand, LEAD retrieves the value of a column from the next row in a result set. You can use LEAD to get the sales value for the next month, like so:

```
SELECT
    Month,
    Sales,
    LEAD(Sales) OVER (ORDER BY Month) AS NextSales
FROM
    SalesTable;
```

This query will return a table with three columns: Month, Sales, and NextSales. The NextSales column shows the sales value from the next month:

Month	Sales	NextSales
1/1/2021	100	150
2/1/2021	150	200
3/1/2021	200	250
4/1/2021	250	300
5/1/2021	300	350
6/1/2021	350	400

7/1/2021	400	450
8/1/2021	450	500
9/1/2021	500	550
10/1/2021	550	600
11/1/2021	600	650
12/1/2021	650	

Figure 12.6 – Output table

Key KPIs

Based on the lead and lag analysis, some KPIs that can be generated for this dataset include the following:

- **Monthly sales growth rate**: By calculating the percentage change in sales from the previous month, you can track the monthly growth rate and identify trends

- **Average monthly sales**: By taking the average sales over a period of time, such as a quarter or year, you can see how the company is performing overall

- **Seasonality index**: By calculating the average sales for each month over a period of several years, you can identify any seasonal patterns in the data and adjust your strategies accordingly

- **Sales forecast**: By using historical data and applying statistical techniques, you can forecast future sales and plan for future growth

- **Lead time**: By analyzing the lead time between order placement and delivery, you can identify areas where the company can improve its operations and customer satisfaction

- **Inventory turnover**: By dividing the average inventory value by the cost of goods sold, you can calculate how quickly the company is selling its inventory and identify opportunities to optimize inventory levels

These KPIs can help the company make data-driven decisions and track its performance over time.

Percentage change

Calculating percentage change over time is an important analysis technique in time series analysis because it allows us to track the growth rate of a variable (such as sales, revenue, or profit) over time. By calculating the percentage change from one time period to another, we can identify trends, patterns, and anomalies in the data. For example, if we see a consistent positive percentage change in sales over time, we can conclude that our business is growing. On the other hand, if we see a consistent negative percent change, we may need to adjust our business strategies or identify potential issues affecting our sales. Moreover, calculating percentage change allows us to compare the relative change between two time periods, regardless of the absolute level of the variable being measured. This makes it easier to compare and analyze trends across different time periods, which is especially useful for identifying seasonality or cyclical patterns in the data. Overall, by calculating percentage change over time, we can gain valuable insights into the performance of our business, make informed decisions, and drive growth and profitability.

Case scenario

We can calculate percent change using the previous dataset in SQL for a time series.

Assuming the sales data is stored in a table called `sales_data` with `date` and `sales` columns, we can calculate the monthly percent change in sales using the following SQL query:

```
SELECT
    date,
    sales,
    (sales - LAG(sales) OVER (ORDER BY date)) / LAG(sales) OVER (ORDER
BY date) AS monthly_percent_change
FROM
    sales_data;
```

The `LAG()` function is used to access the previous month's sales data, and then the percent change in sales between the current month and the previous month is calculated by dividing the difference by the previous month's sales.

This query will return a table with columns for the date, sales, and monthly percent change in sales. By calculating percent changes over time, we can track the growth rate and identify trends in the sales data:

date	sales	monthly_percent_change
1/1/2021	100	NULL
2/1/2021	150	50
3/1/2021	200	33.3
4/1/2021	250	25
5/1/2021	300	20
6/1/2021	350	16.7
7/1/2021	400	14.3
8/1/2021	450	12.5
9/1/2021	500	11.1
10/1/2021	550	10
11/1/2021	600	9.1
12/1/2021	650	8.3

Figure 12.7 – Output table

Key KPIs

Using the SQL function to calculate monthly percent change, we can derive the following KPIs:

- **Monthly sales growth rate**: By calculating the monthly percent change in sales, we can track the growth rate and identify trends in the data

- **Average monthly percent change**: By calculating the average of the monthly percent changes over a specific period of time, we can gain insights into the overall trend in sales growth

- **Seasonality index**: By calculating the average monthly percent change for each month over a period of several years, we can identify any seasonal patterns in the data and adjust our strategies accordingly

- **Sales forecast**: By using historical data and applying statistical techniques, we can forecast future sales and plan for future growth

- **Sales volatility**: By calculating the standard deviation of the monthly percent changes, we can measure the volatility of sales and identify any outliers or abnormal changes in the data

- **Sales momentum**: By calculating the moving average of the monthly percent changes, we can identify trends in the data and measure the momentum of sales growth

These KPIs can help us track our sales performance, identify areas for improvement, and make data-driven decisions to drive growth and profitability.

Moving averages

A moving average is a time series analysis technique that is commonly used to smooth out fluctuations in the data and identify trends or patterns. In SQL, a moving average can be calculated using a window function. A moving average is the average of a fixed number of the most recent data points in a time series. For example, a 3-month moving average for monthly sales data would be the average of the most recent 3 months of sales data. As new data becomes available, the moving average "moves" forward, with the oldest data point being dropped from the calculation and the newest data point being added. A moving average is important because it helps to remove noise or fluctuations in the data and provides a clearer view of underlying trends or patterns. This is especially useful in identifying seasonal or cyclical patterns, as well as longer-term trends in the data. Moreover, moving averages can also help to identify turning points or changes in the direction of the data. For example, if the moving average has been consistently increasing over time but suddenly starts to decrease, this could indicate a shift in trend or a change in the underlying factors affecting the data.

Overall, a moving average is a powerful time series analysis technique that can help to uncover important insights and trends in the data. By using SQL to calculate moving averages, we can perform this analysis quickly and efficiently and make informed decisions based on the results.

Case scenario

A moving average is a powerful time series analysis technique that can be particularly useful in tracking the progression of the COVID-19 pandemic across different countries. By calculating the moving average of daily COVID-19 cases, we can identify trends and patterns in the data and make informed decisions about public health policy and resource allocation.

For example, let's consider the COVID-19 case data for a particular country. The daily case count may be affected by various factors, such as the availability of testing, reporting delays, and changes in public behavior. As a result, the raw data may contain a lot of noise and fluctuations, making it difficult to identify underlying trends.

By calculating the moving average of daily COVID-19 cases, we can smooth out noise in the data and provide a clearer view of the underlying trends. This can help us to identify the peaks and troughs of the pandemic and track how the situation is evolving over time. We can also use moving averages to compare the progression of the pandemic across different countries and identify areas that may require more resources or attention.

Moreover, by analyzing the trend of the moving average over time, we can identify changes in the growth rate of the pandemic and adjust our public health policies accordingly. For example, if the moving average has been increasing rapidly, we may need to implement stricter measures such as lockdowns or mandatory mask-wearing. Conversely, if the moving average has been decreasing, we may consider easing some of these measures. Overall, a moving average is a powerful time series analysis technique that can help us to make informed decisions about public health policy and resource allocation in the fight against the COVID-19 pandemic.

Date	Daily_Cases
1/1/2022	1000
1/2/2022	1200
1/3/2022	900
1/4/2022	1500
1/5/2022	1300
1/6/2022	1800
1/7/2022	2000
1/8/2022	1900
1/9/2022	2200
1/10/2022	2400

Figure 12.8 – Covid cases table

We can calculate the moving average of daily COVID-19 cases over a certain period of time using the following SQL query:

```
SELECT Date, Daily_Cases, AVG(Daily_Cases) OVER (ORDER BY Date ROWS
BETWEEN 4 PRECEDING AND CURRENT ROW) AS 5_Day_Moving_Average
FROM covid_cases
ORDER BY Date;
```

This query uses the AVG function along with the OVER clause to calculate the moving average of daily cases. The ROWS BETWEEN 4 PRECEDING AND CURRENT ROW part of the query specifies that the moving average should be calculated over a period of 5 days (including the current day).

The output of this query would be as follows:

Date	Daily_Cases	5_Day_Moving_Average
1/1/2022	1000	1000
1/2/2022	1200	1100
1/3/2022	900	1033.33
1/4/2022	1500	1150
1/5/2022	1300	1180
1/6/2022	1800	1360
1/7/2022	2000	1560
1/8/2022	1900	1760
1/9/2022	2200	1960
1/10/2022	2400	2040

Figure 12.9 – Output table

As we can see, the moving average smooths out noise in the data and provides a clearer view of underlying trends. We can use this information to identify the peaks and troughs of the pandemic, track how the situation is evolving over time, and make informed decisions about public health policy and resource allocation.

Key KPIs

Moving averages can be used to generate several KPIs for time series data. Some common KPIs include the following:

- **Trend analysis**: Moving averages can help identify the overall trend of time series data by smoothing out noise and fluctuations in the data.

- **Seasonality analysis**: By analyzing the moving averages over different time periods, seasonality patterns can be identified. This can help businesses in planning for seasonal changes in demand, pricing, and marketing strategies.

- **Forecasting**: Moving averages can be used for short-term forecasting by analyzing trends and seasonality patterns in the data.

- **Performance tracking**: By comparing moving averages over different time periods, businesses can track their performance and identify any anomalies or changes in the data.

- **Volatility analysis**: Moving averages can help identify volatility in the data by measuring the spread of moving average values around the mean value.

- **Momentum analysis**: By calculating the difference between moving averages over different time periods, momentum can be analyzed. This can help identify trends and predict changes in the data.

Overall, moving averages are a powerful tool for analyzing time series data and can provide valuable insights for businesses to make informed decisions.

Rank for time series analysis

A rank is a measure used in time series analysis to determine the relative position of a data point within a set of historical observations. In other words, it helps you understand how a particular data point compares to other data points in terms of its magnitude or value.

Case scenario

For example, if you have a time series of daily temperature readings over the course of a month, you might use a rank to determine whether a particular day's temperature was unusually high or low compared to the rest of the month.

To calculate the rank, you would simply order all of the data points in your time series from lowest to highest (or vice versa) and assign each point a rank based on its position in the ordering. The highest value would receive a rank of 1, the second-highest a rank of 2, and so on.

Once you have calculated the rank for a particular data point, you can compare it to the ranks of other data points to gain insights into how it fits into the overall pattern of the time series. For example, if a data point has a very low rank (that is, it is one of the smallest values in the time series), it might indicate that something unusual or anomalous is happening in the data. Conversely, a data point with a very high rank (that is, it is one of the largest values in the time series) might suggest that a trend or pattern is driving the data in a particular direction.

Let's say we have the following raw time series dataset, which shows the daily temperature readings for the first 10 days of April:

Day	Temperature
1	15
2	18
3	16
4	14
5	19
6	22
7	21
8	20
9	17
10	16

Figure 12.10 – Time series dataset

To calculate the rank of each temperature reading, we would first sort the dataset in ascending order by temperature. When two or more values are equal, they receive the same rank, which is the average of the ranks that would have been assigned if they were unique. So, in this case, we would get the following table:

Day	Temperature	Rank
4	14	1
1	15	2.5
3	16	4.5
10	16	4.5
9	17	6
8	20	7
7	21	8
2	18	9
5	19	10
6	22	11

Figure 12.11 – Time series dataset with rank function

As you can see, the temperature reading of 14 on day 4 has the lowest value and therefore receives a rank of 1. The highest value is 22 on day 6 and receives a rank of 11. Now, we can use the rank to gain insights into the data. For example, we can see that the temperature on day 4 is the lowest in the dataset, as it has the lowest rank of 1. We can also see that there is a tie for the second-lowest temperature, with both days 1 and 3 receiving a rank of 2.5.

Ranks can be useful for comparing the relative position of data points over time and can help identify trends or anomalies in the data.

Key KPIs

Several KPIs can be generated using rank analysis in time series data. Here are a few examples:

- **Minimum and maximum values**: The minimum and maximum values in a time series can be identified using rank analysis. These values can be used to track performance and identify anomalies in the data.

- **Percentile rankings**: Percentile rankings can be calculated based on the rank of a particular data point. For example, the 90th percentile might be the temperature at which 90% of the observations fall below that value. This can be useful for identifying outliers and setting benchmarks for performance.

- **Trend analysis**: By tracking the rank of a particular data point over time, you can identify trends in the data. For example, if the rank of a particular metric is consistently increasing over time, it might indicate that performance is improving.

- **Seasonality analysis**: Rank analysis can also be used to identify seasonal trends in the data. For example, if the rank of a particular metric is consistently higher during the summer months than during the winter months, it might indicate that there is a seasonal pattern in the data.

Overall, the key KPIs that can be generated using rank analysis depend on the specific time series data and the business objectives. By understanding patterns and trends in the data, businesses can make more informed decisions about performance and strategy.

CTE for time series analysis

A **common table expression** (CTE) is a temporary named result set in SQL that you can use to perform complex queries. In time series analysis, CTEs can be used to create a rolling window of data that you can use to calculate metrics such as moving averages, percentiles, and other statistical measures.

Here is an example of how you could use a CTE in SQL to calculate a rolling average of temperature readings over a 7-day period:

```
WITH rolling_average AS (
  SELECT day, temperature,
        AVG(temperature) OVER (ORDER BY day ROWS BETWEEN 6 PRECEDING
AND CURRENT ROW) AS rolling_avg
  FROM temperature_data
)
SELECT day, temperature, rolling_avg
FROM rolling_average;
```

In this example, we first create a CTE called `rolling_average` that includes the day, temperature, and a rolling average calculated using the `AVG` function and a window that includes the current row and the six preceding rows. The `OVER` clause specifies the ordering of the data by day. This creates a rolling window of 7 days over which the average temperature is calculated.

In the second part of the query, we select the day, temperature, and rolling average from the `rolling_average` CTE. This gives us a table of data that includes the daily temperature readings as well as the rolling average for each day.

Using a CTE in SQL for time series analysis allows you to easily perform calculations that require a rolling window of data. You can use this approach to calculate other statistical measures, such as percentiles, standard deviations, or correlations, to gain insights into trends and patterns in the data.

Importance of using CTEs while performing time series analysis

CTEs are helpful while performing time series analysis for several reasons, as follows:

- **Rolling windows**: CTEs can be used to create rolling windows of data, which are often necessary for time series analysis. Rolling windows allow you to calculate metrics such as moving averages, percentiles, and other statistical measures based on a window of data that moves through time. CTEs can be used to define this window and perform calculations within it.

- **Simplify complex queries**: Time series data can often be complex, with multiple variables and dimensions. CTEs can help simplify complex queries by breaking them down into smaller, more manageable pieces. You can use CTEs to create intermediate tables that can be joined together to create a final output.

- **Reusability**: CTEs can be used multiple times within a query, making them reusable and reducing the need for redundant code. This can save time and effort when working with large and complex datasets.

- **Organizing code**: CTEs can help organize code and make it more readable. By breaking down a complex query into smaller, more manageable pieces, you can improve the readability and maintainability of the code.

Overall, CTEs can be a powerful tool for time series analysis as they allow you to perform complex calculations and manipulations on time series data with ease. They simplify queries, organize code, and improve the readability and maintainability of the code.

Forecasting with linear regression

Forecasting with linear regression in SQL is useful because it allows us to make predictions about future values based on historical data and other relevant factors. By using SQL to build and analyze these models, we can gain insights into how different factors affect our business and make more informed decisions about how to allocate resources and plan for the future.

Some of the benefits of using linear regression for forecasting in SQL include the following:

- **Simplicity**: Linear regression is a simple and widely used technique for predicting future values. By using SQL to build and analyze these models, we can easily incorporate data from different sources and perform calculations that might be difficult or time-consuming in other software.

- **Flexibility**: Linear regression can be used to predict a wide range of values, including sales, website traffic, customer engagement, and more. By using SQL to build and analyze these models, we can tailor our forecasts to our specific business needs.

- **Interpretability**: Linear regression models are relatively easy to interpret, which means that we can gain insights into how different factors affect our business and make more informed decisions about how to allocate resources and plan for the future.

Overall, forecasting with linear regression in SQL is a powerful tool for businesses that want to make data-driven decisions and plan for the future based on historical trends and other relevant factors.

Case scenario

Let's quickly understand how you could use SQL to perform forecasting with linear regression on e-commerce sales data for a company such as FedEx, Amazon, and so on. Let's say you have a `sales_data` table that includes the following columns:

- `order_date`: The date the order was placed
- `revenue`: The total revenue for the order
- `num_shipments`: The total number of shipments for the order

To perform forecasting with linear regression on this data, you would follow these steps:

1. **Gather data**: Retrieve the data you want to use for the regression analysis. In this case, you would retrieve the `order_date`, `revenue`, and `num_shipments` columns from the `sales_data` table, like so:

    ```
    SELECT order_date, revenue, num_shipments
    FROM sales_data;
    ```

2. **Calculate features**: Calculate any additional features you want to include in your regression analysis. For example, you might want to include a `day_of_week` feature to capture any weekly patterns in the sales data, as follows:

```
SELECT
  order_date,
  revenue,
  num_shipments,
  EXTRACT(DOW FROM order_date) AS day_of_week
FROM sales_data;
```

3. **Build model**: Use the `LINEAR_REGR` function in SQL to build a linear regression model using historical data, like so:

```
SELECT
  LINEAR_REGR(revenue, day_of_week) OVER() AS revenue_slope,
  AVG(revenue) - LINEAR_REGR(revenue, day_of_week) OVER() * AVG(day_
of_week) AS revenue_intercept,
  LINEAR_REGR(num_shipments, day_of_week) OVER() AS num_shipments_
slope,
  AVG(num_shipments) - LINEAR_REGR(num_shipments, day_of_week) OVER()
* AVG(day_of_week) AS num_shipments_intercept
FROM sales_data;
```

This will give you the slopes and intercepts of the linear regression lines for revenue and the number of shipments.

4. **Make predictions**: Use the linear regression equations to make predictions about future revenue and the number of shipments. For example, if you want to predict the revenue and number of shipments for the next day, you would use the following query:

```
WITH model_parameters AS (
  SELECT
    LINEAR_REGR(revenue, day_of_week) OVER() AS revenue_slope,
    AVG(revenue) - LINEAR_REGR(revenue, day_of_week) OVER() * AVG(day_
of_week) AS revenue_intercept,
    LINEAR_REGR(num_shipments, day_of_week) OVER() AS num_shipments_
slope,
    AVG(num_shipments) - LINEAR_REGR(num_shipments, day_of_week)
OVER() * AVG(day_of_week) AS num_shipments_intercept
  FROM sales_data
),
predicted_values AS (
  SELECT
    DATE_TRUNC('day', CURRENT_DATE + INTERVAL '1 day') AS prediction_
date,
    (SELECT revenue_slope * EXTRACT(DOW FROM CURRENT_DATE + INTERVAL
'1 day') + revenue_intercept FROM model_parameters) AS predicted_
revenue,
    (SELECT num_shipments_slope * EXTRACT(DOW FROM CURRENT_DATE +
INTERVAL '1 day') + num_shipments_intercept FROM model_parameters) AS
predicted_num_shipments
)
SELECT
  prediction_date,
  ROUND(predicted_revenue, 2) AS predicted_revenue,
  ROUND(predicted_num_shipments, 2) AS predicted_num_shipments
FROM predicted_values;
```

This will give you the predicted revenue and number of shipments for the next day, based on the linear regression model built using historical data. You can adjust the query to predict different time intervals by changing the INTERVAL parameter in the predicted_values subquery.

Key KPIs

Several KPIs can be generated using forecasting with linear regression in SQL. These include the following:

1. **Forecast accuracy**: This KPI measures how well our model is able to predict future values based on historical data. We can measure forecast accuracy using metrics such as **mean absolute error (MAE)**, **mean squared error (MSE)**, or **root mean squared error (RMSE)**.

2. **Sales forecasting**: This KPI measures how well we are able to predict future sales based on historical sales data and other relevant factors. By forecasting sales accurately, we can make more informed decisions about how to allocate resources and plan for the future.

3. **Customer engagement forecasting**: This KPI measures how well we are able to predict future customer engagement based on historical engagement data and other relevant factors. By forecasting customer engagement accurately, we can make more informed decisions about how to target our marketing efforts and improve customer satisfaction.

4. **Inventory forecasting**: This KPI measures how well we are able to predict future inventory levels based on historical inventory data and other relevant factors. By forecasting inventory levels accurately, we can make more informed decisions about how much stock to order and when to place orders.

Overall, the key KPIs generated using forecasting with linear regression in SQL will depend on the specific business needs and goals of the organization. However, accurate forecasting is crucial for making informed decisions and staying ahead of the competition, regardless of the industry or type of organization.

Summary

This brings us to the end of this chapter, where we have learned about how to use SQL to analyze time series data and make predictions about future values based on historical trends and other relevant factors. Specifically, you learned the following:

- How to use SQL to aggregate and visualize time series data, including techniques such as grouping by time intervals, calculating moving averages, and creating line charts and other visualizations

- How to use common time series analysis techniques in SQL, such as calculating seasonality, trend, and volatility, and using CTEs to analyze data over time

- How to use SQL to build and analyze forecasting models, including linear regression models

- How to use SQL to generate KPIs related to time series data, such as forecast accuracy, sales forecasting, customer engagement forecasting, and inventory forecasting

In the next chapter, we will learn different methods to find outliers in the data easily. Outlier detection is an important aspect of data analysis as it helps determine if the data is correct, looks at the skewness of the data, and removes any unexpected values.

13

Outlier Detection

Outlier detection is the process of identifying data points that deviate significantly from the normal behavior of a dataset. **Structured Query Language (SQL)** can be used to perform outlier detection on large datasets. In SQL, outlier detection typically involves analyzing the statistical properties of the dataset, such as the mean, standard deviation, and range. SQL queries can be used to identify data points that fall outside of the expected range of values, or that have a large deviation from the mean. Several methods can be used to perform outlier detection in SQL, including clustering-based methods, distance-based methods, and density-based methods. These methods can be applied to different types of datasets and can be customized based on the specific needs of the analysis. Overall, SQL is a powerful tool for outlier detection in large datasets, allowing analysts to quickly identify and investigate anomalous data points:

Figure 13.1 – Outliers

Some of the reasons why outlier detection is important in data science are as follows:

- **Improving data quality**: Outliers can be caused by data entry errors, measurement errors, or other issues that affect data quality. By identifying and removing outliers, we can improve the overall quality of our data and ensure that our analysis and models are based on accurate information.

- **Enhancing model performance**: Outliers can have a significant impact on the performance of machine learning models. By identifying and removing outliers, we can improve the accuracy and robustness of our models and ensure that they are better able to generalize to new data.

- **Providing insights**: Outliers can sometimes reveal important insights about our data and can help us understand unusual or unexpected phenomena. By analyzing outliers, we can gain new insights into our data and potentially identify new patterns or relationships that we may have otherwise missed.

Overall, outlier detection in SQL is an important tool in any data scientist's toolkit, helping ensure that we are working with high-quality data and producing accurate and meaningful insights.

Measures of central tendency and dispersion

This section covers measures of central tendency, such as mean, median, and mode, and measures of dispersion, such as range, variance, and standard deviation. Let's understand this with the help of an example scenario.

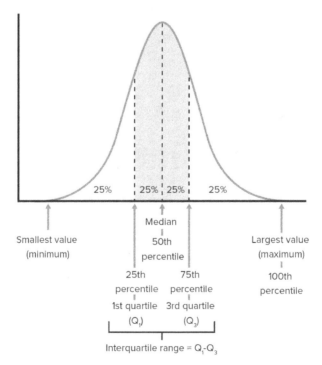

Figure 13.2 – Central tendency and dispersion

Case scenario

Suppose we have a group of people who work in a company, and we want to understand how much they are paid. We can calculate the average salary for the group by adding up all their salaries and dividing by the number of people. This gives us an idea of how much money they typically earn. However, there may be some people in the group who earn much more or much less than the others. These are called outliers, and they can skew the average salary and make it less representative of the group as a whole. To identify outliers, we can use a statistical measure called the standard deviation. The standard deviation tells us how spread out the salaries are from the average. If the salaries are tightly clustered around the average, the standard deviation is small. If the salaries are spread out over a wide range, the standard deviation is large. To detect outliers, we can look for salaries that are significantly higher or lower than the average salary by a certain amount, which is measured in standard deviations. For example, we might say that any salary that is more than two standard deviations away from the average is an outlier. In SQL, we can write queries to calculate the average salary, standard deviation, and z-score (which is a measure of how far a salary is from the average in terms of standard deviations) for the group of employees. Then, we can use conditional statements to identify the outliers based on their z-scores.

Let's use a raw dataset to understand this.

Suppose we have a table called `employee_salary` in our SQL database that contains the salaries of the employees in a company:

Employee ID	Salary
1	50000
2	55000
3	60000
4	65000
5	70000
6	75000
7	80000
8	85000
9	90000
10	95000
11	100000
12	200000

Figure 13.3 – employee_salary

Our goal is to identify any outliers in the dataset using the standard deviation method in SQL. Here are the steps we can follow:

1. Calculate the mean and standard deviation of the salaries using the built-in `AVG()` and `STDEV()` SQL functions:

   ```
   SELECT AVG(salary), STDEV(salary) FROM employee_salary;
   ```

   This query returns a mean salary of **$78,846** and a standard deviation of **$33,048**.

2. Calculate the z-score for each salary using the (salary - mean) / standard deviation formula:

   ```
   SELECT employee_id, salary, (salary - 78846) / 33048 AS z_score
   FROM employee_salary;
   ```

   This query returns a table stating the z-score for each salary:

Employee ID	Salary	Z-Score
1	50000	-0.98
2	55000	-0.62
3	60000	-0.25
4	65000	0.12
5	70000	0.49
6	75000	0.86
7	80000	1.23
8	85000	1.6
9	90000	1.97
10	95000	2.34
11	100000	2.71
12	200000	4.81

Figure 13.4 – Z-score of each salary

- Identify any salaries with a z-score greater than a certain threshold, which indicates that they are outliers.

A common threshold is two, meaning any salary that is more than two standard deviations away from the mean is an outlier. However, it's important to note that the choice of the threshold may vary, depending on the context, the characteristics of the dataset, and the requirements of the analysis. In some cases, a more stringent threshold, such as a Z-score of three or higher, may be used to identify outliers. It's essential to consider the distribution and nature of the data before determining the appropriate threshold for identifying outliers based on the Z-score:

```
SELECT employee_id, salary, (salary - 78846) / 33048 AS z_score
FROM employee_salary
WHERE (salary - 78846) / 33048 > 2 OR (salary - 78846) / 33048 < -2;
```

This query returns a table with any salaries that are more than two standard deviations away from the mean:

Employee ID	Salary	Z-Score
12	200000	4.81

Figure 13.5 – Salaries beyond two standard deviations

In this example, we can see that employee 12 has a salary that is more than four standard deviations away from the mean, making it a clear outlier.

Key KPIs

Several **key performance indicators** (**KPIs**) can be generated using outlier analysis with standard deviation in SQL:

- **The number of outliers**: This KPI measures the total number of outliers identified in the dataset. It can indicate how significant the outliers are and how much they are affecting the overall distribution of the data.

- **The percentage of outliers**: This KPI measures the proportion of data points that are outliers, expressed as a percentage of the total dataset. This can give you an idea of how much the outliers are deviating from the normal distribution of the data.

- **The average deviation from the mean**: This KPI measures the average distance of the outliers from the mean of the dataset. This can give you an idea of how far the outliers are from the normal distribution and how much they are affecting the overall distribution.

- **Impact of outliers on statistical measures**: This KPI measures how much the outliers are affecting the statistical measures of the dataset, such as the mean, median, and standard deviation. By comparing the measures with and without outliers, it is possible to understand how much impact the outliers are having on the overall analysis.

- **Distribution of outliers**: This KPI measures the distribution of the outliers across different categories or dimensions of the dataset. For example, it may be possible to identify which departments or locations have the highest number of outliers, which can help you identify potential issues or areas for improvement.

Methods for detecting outliers

This section covers different methods for detecting outliers, such as the z-score, box plots, and density-based methods.

Let's consider a dataset of customer orders that contains the `order_id`, `customer_id`, `product_name`, and `price` fields. We will create a table called `orders` so that we can store this data:

```
CREATE TABLE orders (
    order_id INT,
    customer_id INT,
    product_name VARCHAR(255),
    price FLOAT
);

INSERT INTO orders (order_id, customer_id, product_name, price)
VALUES
(1, 101, 'Product A', 10.50),
(2, 102, 'Product B', 20.25),
(3, 103, 'Product C', 12.75),
(4, 104, 'Product A', 9.50),
(5, 105, 'Product B', 21.00),
(6, 106, 'Product D', 8.25),
(7, 107, 'Product E', 15.75),
(8, 108, 'Product B', 22.50),
(9, 109, 'Product A', 11.25),
(10, 110, 'Product F', 7.50);
```

Box plot method

Let's use the box plot method to identify outliers based on the `price` field. We can use the following query to calculate the quartiles and 95th percentile of the `price` column for each product name:

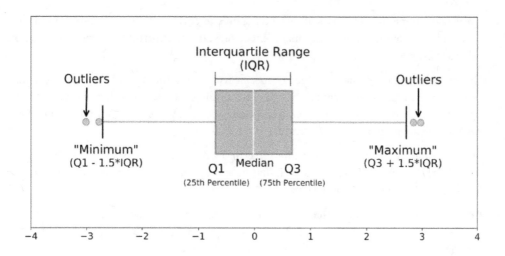

Figure 13.6 – Different parts of a box plot

```
SELECT product_name,
   PERCENTILE_CONT(0.25) WITHIN GROUP (ORDER BY price) AS q1,
   PERCENTILE_CONT(0.5) WITHIN GROUP (ORDER BY price) AS median,
   PERCENTILE_CONT(0.75) WITHIN GROUP (ORDER BY price) AS q3,
   PERCENTILE_CONT(0.95) WITHIN GROUP (ORDER BY price) AS q95
FROM orders
GROUP BY product_name;
```

This query returns the quartiles and 95th percentile of the price column for each product name. Any data points with a price greater than the upper whisker (q3 + 1.5 * IQR) or less than the lower whisker (q1 - 1.5 * IQR) are considered outliers.

Handling outliers

Handling outliers in SQL involves deciding what to do with the data points that have been identified as outliers using outlier detection methods.

There are several ways to handle outliers in SQL, depending on the context and the nature of the data. Here are some common approaches:

- **Remove outliers**: One option is to simply remove the outliers from the dataset. This can be appropriate if the outliers are due to data entry errors or other anomalies that are unlikely to occur again. You can use SQL queries to filter out the outlier data points from the dataset.

- **Correct outliers**: In some cases, it may be possible to correct the outliers. For example, if a data point is an outlier due to a decimal point error, you could correct the error and adjust the value accordingly.

- **Group outliers**: If the outliers are due to a specific reason, you could group them and treat them as a separate category. For example, if you are analyzing product sales data and you identify a group of high-value purchases from a corporate account, you could group these purchases and analyze them separately from the rest of the data.

- **Ignore outliers**: In some cases, the outliers may be legitimate data points that are truly different from the rest of the data. In such cases, it may be appropriate to simply ignore the outliers and focus on analyzing the rest of the data.

- **Investigate outliers**: If the outliers are not easy to explain, it may be necessary to investigate them further to determine the cause. This could involve looking at additional data sources or conducting further analysis to understand why the outliers occurred.

Overall, handling outliers in SQL requires carefully considering the data and the context in which it was collected. By using appropriate techniques to handle outliers, you can ensure that your analysis is accurate and reliable and that your conclusions are based on sound data.

Case scenario

Let's consider a real-world practical scenario where handling outliers in SQL is important.

Suppose you are working for a financial institution and you are analyzing loan data. Your dataset includes various features, such as loan amount, borrower income, credit score, and so on. You want to identify any potential outliers in the data to ensure that the loans you are approving are legitimate and financially sound.

Using outlier detection methods in SQL, you identify a set of loan applications that have extremely high loan amounts compared to the rest of the data. You need to decide how to handle these outliers to ensure that your loan portfolio is not at risk.

One option is to remove the outliers from the dataset. However, upon further investigation, you find that the high loan amounts are due to the borrowers taking out loans for commercial real estate projects. These loans are legitimate and financially sound, but they are significantly larger than the typical consumer loans in the dataset. In this case, it may be appropriate to group the outliers and treat them as a separate category. By doing so, you can analyze them separately from the rest of the data and ensure that your loan portfolio is balanced and financially sound.

Another option is to investigate the outliers further to determine if they are due to data entry errors or other anomalies. If you find that the outliers are due to errors, you can correct them or remove them from the dataset. If they are legitimate data points, you may need to adjust your analysis to account for their impact on the overall data.

Key points to keep in mind while handling outliers

Here are some key points to keep in mind:

- **Understand the nature of the data**: Before handling outliers, it's important to understand the nature of the data and how outliers can impact your analysis. Are the outliers due to data entry errors or do they represent legitimate data points? Are the outliers important to your analysis or can they be safely removed?

- **Choose an appropriate technique**: There are various techniques to handle outliers in SQL, such as removing them, adjusting their values, or treating them as a separate category. It's important to choose the appropriate technique based on the nature of the data and the goals of your analysis.

- **Document your approach**: It's important to document your approach to handling outliers in SQL so that others can understand your methodology and replicate your analysis. This documentation should include details about the technique used, the rationale for choosing that technique, and any assumptions or limitations.

- **Check for unintended consequences**: Handling outliers can have unintended consequences on your analysis, such as skewing the distribution of the data or introducing bias. It's important to check for these unintended consequences and adjust your analysis accordingly.

- **Re-evaluate your analysis**: After handling outliers, it's important to re-evaluate your analysis to ensure that your conclusions are still valid. This may involve rerunning statistical tests or adjusting your visualizations to account for the outliers.

By keeping these key points in mind, you can ensure that your approach to handling outliers in SQL is robust and produces reliable results.

Applying outlier detection

Outlier detection using SQL can be applied in various industries and use cases where identifying and addressing unusual data points is important. Here are some examples:

- **Fraud detection**: SQL can be used to identify unusual transactions or behaviors in financial data that may indicate fraud. By using SQL to analyze large volumes of transaction data, financial institutions can identify potential fraudsters and take action to prevent fraudulent activity.

- **Healthcare**: SQL can be used to identify unusual patient health data, such as abnormal lab test results, which may indicate the presence of a disease or health condition. By using SQL to analyze patient health data in real time, healthcare providers can identify potential health issues and provide early intervention.

- **Retail**: SQL can be used to identify unusual customer behavior, such as high-value purchases or returns, which may indicate potential fraud or theft. By using SQL to analyze customer transaction data, retailers can identify potential fraudsters and take action to prevent further fraudulent activity.

- **Marketing**: SQL can be used to identify unusual patterns in customer behavior, such as a sudden spike in website traffic or social media engagement, which may indicate the effectiveness of a marketing campaign or the presence of a new trend. By using SQL to analyze customer data in real time, marketers can adjust their strategies and improve their campaigns.

- **Energy management**: SQL can be used to identify unusual patterns in energy consumption data, which may indicate energy waste or equipment malfunction. By using SQL to analyze energy consumption data in real time, energy providers can identify areas for improvement and optimize their operations.

Overall, outlier detection using SQL is a valuable tool for identifying unusual data patterns in various industries and use cases. By using SQL to analyze data in real time, organizations can identify potential issues and take proactive action to address them.

Case scenario

Let's see how outlier detection using SQL can be applied in manufacturing:

- **Quality control**: SQL can be used to identify outliers in product quality data, such as measurements of product dimensions, weight, or other quality metrics. By using SQL to filter outlying data points, manufacturers can identify defective or nonconforming products and take corrective action to improve overall product quality.

- **Equipment maintenance**: SQL can be used to monitor equipment performance data, such as temperature, vibration, or other sensor readings, and identify any anomalies or outliers that may indicate impending equipment failure. By using SQL to analyze this data in real time, manufacturers can schedule maintenance or repairs proactively before the equipment fails and causes costly downtime.

- **Supply chain management**: SQL can be used to identify outliers in supply chain data, such as lead times, delivery schedules, or quality metrics for raw materials or components. By using SQL to analyze this data, manufacturers can identify suppliers or vendors who consistently provide products or components that fall outside of expected quality standards, and work with them to improve product quality and reduce the risk of supply chain disruptions.

- **Predictive maintenance**: SQL can be used to develop predictive maintenance models that use historical equipment performance data to identify outliers and predict equipment failures before they occur. By using SQL to analyze this data and build predictive models, manufacturers can schedule maintenance or repairs proactively, before the equipment fails and causes costly downtime.

- **Process optimization**: SQL can be used to identify outliers in production process data, such as cycle times, inventory levels, or other performance metrics. By using SQL to analyze this data, manufacturers can identify inefficiencies or bottlenecks in their processes and optimize them to improve efficiency, reduce waste, and increase overall productivity.

Overall, outlier detection using SQL is a valuable tool for manufacturers looking to improve their operations, increase efficiency, and reduce costs. By using SQL to identify and address outliers in real time, manufacturers can improve product quality, reduce downtime, and improve overall customer satisfaction.

Challenges and limitations

In this section, we will discuss the challenges and limitations associated with outlier detection in SQL, such as data quality issues, algorithmic complexity, and computational resources:

- **Data quality issues**: Outlier detection algorithms are sensitive to the quality of the data being analyzed. If the data is incomplete, contains errors, or is inconsistent, it can lead to inaccurate results. Therefore, it is important to ensure that the data being analyzed is of high quality.

- **Algorithmic complexity**: Some outlier detection algorithms can be computationally expensive, which means that they can take a long time to run on large datasets. This can be a challenge for organizations that need to process data in real time.

- **Computational resources**: Outlier detection algorithms require a significant amount of computational resources, such as processing power and memory. This can be a challenge for organizations that do not have access to these resources or that need to process large amounts of data.

- **Interpretability of results**: Outlier detection algorithms can provide results that are difficult to interpret. For example, it may not be clear why a particular data point has been identified as an outlier or what action should be taken in response.

- **Algorithm selection**: There are many different outlier detection algorithms to choose from, each with its own strengths and weaknesses. Selecting the appropriate algorithm for a particular dataset can be a challenge, and it may require some trial and error.

Overall, outlier detection in SQL is a powerful tool, but it does come with some challenges and limitations. It is important to understand these limitations and to use outlier detection in conjunction with other data analysis techniques to ensure accurate results.

Best practices

This section provides recommendations for best practices in outlier detection in SQL, such as selecting appropriate methods based on the data's characteristics, validating the results, and documenting the process.

Here are some best practices for outlier detection in SQL:

- **Select the appropriate methods**: It is important to select the appropriate outlier detection method based on the data's characteristics.

  - **Unskewed dataset**: If we have data that follows a bell-shaped curve (like the shape of a hill), which is called a Gaussian distribution or normal distribution, we can use certain techniques to analyze and make sense of that data. One technique is called the Z-score. It helps us understand how far away a particular data point is from the average (mean) value of the data. Imagine a group of students taking a test. The average score is the expected or typical score. The Z-score tells us how many standard deviations a student's score is above or below the average. A positive Z-score means the score is above average, while a negative Z-score means it's below average. This helps us identify outliers, which are scores that are significantly higher or lower than the average. Another technique is called the Mahalanobis distance. It helps us measure the similarity or dissimilarity between data points based on multiple variables. Imagine a group of people described by their height, weight, and age. The Mahalanobis distance takes into account the relationships between these variables and calculates a distance metric. It helps us identify data points that are unusual or different from the rest of the group. For example, if most people in the group have a similar height, weight, and age, a data point with a high Mahalanobis distance might indicate someone very different from the others.

  - **Skewed dataset**: When we talk about skewed data or heavy tails, we're referring to situations where the data doesn't follow a symmetric bell-shaped curve. Instead, it might be stretched out to one side or have some extreme values that deviate from the majority of the data. In such cases, there are different techniques we can use to understand and analyze the data. One technique is the box plot. Imagine we have a set of data representing people's ages. The box plot helps us visualize the distribution of the ages and identify any outliers. It displays a rectangular box that represents the middle 50% of the data. The line inside the box represents the median age (the middle value). The "whiskers" extend from the box to show the range of the data, excluding outliers. Outliers are data points that fall significantly outside the range of the rest of the data. By looking at this box plot, we can quickly see if any unusual or extreme values might affect our analysis. Another technique is using percentiles. Percentiles divide the data into equal parts. For example, the 25th percentile represents the value below which 25% of the data falls. Similarly, the 75th percentile represents the value below which 75% of the data falls. Percentiles help us understand the distribution of the data and identify specific cutoff points. For instance, if we're looking at students' test scores, we can see how many students scored above or below a particular percentile, such as the 90th percentile. This helps us compare individual scores to the overall performance of the group. In summary, when dealing with data that is skewed or has heavy tails (meaning it deviates from a symmetric pattern), techniques such as box plots and percentiles are useful.

- **Validate the results**: Once an outlier detection method has been applied, it is important to validate the results. This can be done by comparing the results to domain knowledge, verifying that the detected outliers are truly anomalous, and testing the robustness of the method to different data subsets.

- **Document the process**: It is important to document the entire outlier detection process, including the data preprocessing steps, selecting the outlier detection method, and validating the results. This documentation should include the rationale for each step in the process, the code used, and the results obtained.

- **Remove or handle outliers**: Once the outliers have been detected, it is important to decide how to handle them. Depending on the context, outliers may need to be removed from the dataset, or they may need to be retained and treated as a separate category. It is important to carefully consider the implications of each approach and to document the decision-making process.

Overall, following these best practices can help ensure that the outlier detection process is robust, transparent, and reproducible and that the results are accurate and reliable.

Summary

This brings us to the end of this chapter, where we learned what outliers are, measures of central tendency and dispersion, and methods for detecting outliers. We covered code examples for implementation, strategies for handling outliers, and applications of outlier detection in various fields, including their challenges and limitations. We also considered the best practices for selecting appropriate methods based on data characteristics, validating the results, and documenting the process.

Index

P

percentage change 288
  case scenario 289
  key KPIs 290
percentile function 222
PIVOT function
  use case scenario 144-147
  using 142, 143
pivoting data, in SQL 141, 142
  data, transposing from rows
    to columns 142, 143
primary key 12, 14
Publishing step, data wrangling 37
  exercise 57

Q

quarter-over-quarter (Q-O-Q) 28
query
  database monitoring 265
  logging 263-265
  monitoring 261
  profiling 262, 263
  tips and tricks, for writing 266
  troubleshooting 261
query execution plans 256
  limitations, addressing 256
query optimization 247, 255
  better decision-making 256
  better performance, for business-
    critical queries 256
  cost savings 256
  improved accuracy 256
  resource utilization 256
  scalability 256
  user satisfaction 256

query optimization techniques 257
  aggregation 258, 259
  caching 260, 261
  example scenario 258, 259
  indexing 257-259
  joins 257
  normalization 261
  query structure 258, 259
  table partitioning 260

R

rank 293
  case scenario 293, 294
  key KPIs 295
RANK() function 238, 239
  scenario 240-243
  versus DENSE_RANK() function 240
ranking functions 221
ratio values 63
  approximate data types 63
  exact data types 63
relational database 6, 7
relational key 11
REPLACE() function 74
  example 75, 76
  parameters 74
REVERSE() function 76
  characteristics 76
RIGHT() function 68
  example 69
root mean squared error (RMSE) 299
ROW_NUMBER() function 233
  example 233, 234
  for eliminating duplicates 237, 238
  purposes 233
  scenario 234-236
R packages 46, 47

Packtpub.com

Subscribe to our online digital library for full access to over 7,000 books and videos, as well as industry leading tools to help you plan your personal development and advance your career. For more information, please visit our website.

Why subscribe?

- Spend less time learning and more time coding with practical eBooks and Videos from over 4,000 industry professionals

- Improve your learning with Skill Plans built especially for you

- Get a free eBook or video every month

- Fully searchable for easy access to vital information

- Copy and paste, print, and bookmark content

Did you know that Packt offers eBook versions of every book published, with PDF and ePub files available? You can upgrade to the eBook version at packtpub.com and as a print book customer, you are entitled to a discount on the eBook copy. Get in touch with us at customercare@packtpub.com for more details.

At www.packtpub.com, you can also read a collection of free technical articles, sign up for a range of free newsletters, and receive exclusive discounts and offers on Packt books and eBooks.

Other Books You May Enjoy

If you enjoyed this book, you may be interested in these other books by Packt:

SQL Query Design Patterns and Best Practices

Steve Hughes, Dennis Neer, Dr. Ram Babu Singh, Shabbir H. Mala, Leslie Andrews, Chi Zhang

ISBN: 9781837633289

- Build efficient queries by reducing the data being returned
- Manipulate your data and format it for easier consumption
- Form common table expressions and window functions to solve complex business issues
- Understand the impact of SQL security on your results
- Understand and use query plans to optimize your queries
- Understand the impact of indexes on your query performance and design
- Work with data lake data and JSON in SQL queries
- Organize your queries using Jupyter notebooks

Data Ingestion with Python Cookbook

Gláucia Esppenchutz

ISBN: 9781837632602

- Implement data observability using monitoring tools
- Automate your data ingestion pipeline
- Read analytical and partitioned data, whether schema or non-schema based
- Debug and prevent data loss through efficient data monitoring and logging
- Establish data access policies using a data governance framework
- Construct a data orchestration framework to improve data quality

Packt is searching for authors like you

If you're interested in becoming an author for Packt, please visit `authors.packtpub.com` and apply today. We have worked with thousands of developers and tech professionals, just like you, to help them share their insight with the global tech community. You can make a general application, apply for a specific hot topic that we are recruiting an author for, or submit your own idea.

Share Your Thoughts

Now you've finished *Data Wrangling with SQL*, we'd love to hear your thoughts! Scan the QR code below to go straight to the Amazon review page for this book and share your feedback or leave a review on the site that you purchased it from.

https://packt.link/r/1-837-63002-X

Your review is important to us and the tech community and will help us make sure we're delivering excellent quality content.

Download a free PDF copy of this book

Thanks for purchasing this book!

Do you like to read on the go but are unable to carry your print books everywhere? Is your eBook purchase not compatible with the device of your choice?

Don't worry, now with every Packt book you get a DRM-free PDF version of that book at no cost.

Read anywhere, any place, on any device. Search, copy, and paste code from your favorite technical books directly into your application.

The perks don't stop there, you can get exclusive access to discounts, newsletters, and great free content in your inbox daily

Follow these simple steps to get the benefits:

1. Scan the QR code or visit the link below

https://packt.link/free-ebook/9781837630028

1. Submit your proof of purchase
2. That's it! We'll send your free PDF and other benefits to your email directly